Saints in Exile

SAINTS IN EXILE

The Holiness-Pentecostal Experience in
African American Religion and Culture

CHERYL J. SANDERS

New York   Oxford
OXFORD UNIVERSITY PRESS
1996

Oxford University Press

Oxford   New York
Athens   Auckland   Bangkok
Calcutta   Cape Town   Dar es Salaam   Delhi
Florence   Hong Kong   Istanbul   Karachi
Kuala Lumpur   Madras   Madrid   Melbourne
Mexico City   Nairobi   Paris   Singapore
Taipei   Tokyo   Toronto

and associated companies in
Berlin   Ibadan

Copyright © 1996 by Cheryl J. Sanders

Published by Oxford University Press, Inc.
198 Madison Avenue, New York, New York 10016

Oxford is a registered trademark of Oxford University Press, Inc.

Library of Congress Cataloging-in-Publication Data
Sanders, Cheryl Jeanne.
Saints in exile : the Holiness-Pentecostal experience in African
American religion and culture / Cheryl J. Sanders.
p.   cm.   — (Religion in America)
Includes index.
ISBN 0-19-509843-9
1. Afro-American Holiness church members.   2. Afro-American
Pentecostals.   3. Church of God (Anderson, Ind.)   4. United States—
Church history—20th century.   I. Title.   II. Series: Religion in
America (New York, N.Y.)
BX7990.H615S26   1996
289.9′4—dc20   95-15690

1 3 5 7 9 8 6 4 2

Printed in the United States of America
on acid-free paper

To the memory of my pastor and mentor
Samuel George Hines
(1929–1995)
Ambassador for Christ, Agent of Reconciliation

Preface

Unlike most scholars who have studied and written about the Sanctified church, I present this study as an "insider." My two children and I represent the third and fourth generations of both my mother's and father's sides of the family in the Church of God (Anderson, Indiana). As far back as I can remember, my immediate family's teachings, beliefs, behavior, and values have been distinctively characteristic of the Holiness tradition. What I would now describe as exilic existence and consciousness colors my earliest memories. My father served in World War II as a conscientious objector because our church was opposed to bearing arms. My mother worked for many years as a social worker and public school administrator, bringing to her profession a level of compassion, concern, and integrity that mirrored her work in the church as a Christian educator.

I recall being surrounded by godly, loving people at church who called themselves "saints." As soon as I became old enough to have an awareness of how most of our neighbors, relatives, and other families lived, I realized that my family was different—my parents did not use alcohol, nicotine, or other addictive drugs, nor did they use profane language, dishonor their marriage vows, cheat on their income taxes, or go to house parties, nightclubs, or movies. Their code of personal ethics was affirmed, by teaching and practice, at our church. Although the specifics of prohibited behaviors have changed in recent years, the saints were people committed to letting their Christian testimony find public manifestation in an ascetic lifestyle and high moral standards.

I have met many people who resented having been brought up in households such as mine and who rebelled as children, adolescents, or adults by engaging in the forbidden behaviors, either secretly or in the open. I made my response to the pastor's weekly altar appeal one Sunday when I was seven years old. After confessing salvation, I found that every time I was tempted to do something I regarded as wrong, the workings of my conscience were influenced by teachings and warnings I had received at church and at home. I felt bad whenever I succumbed to a temptation, and I sought repentance, but I felt affirmed whenever I "escaped." This moral environment aided my transition through childhood and adolescence to adulthood because I often observed the consequences of rebellion and experimentation among those my own age and older, who provided tacit object lessons for me in moral decision making. This is not to say that everyone in my church and family circle was morally perfect but rather that we collectively made clear delineations between right and wrong, especially for the benefit of the young. The do's and don'ts usually made sense to me, especially when they were cited with both biblical and folk injunctions.

I maintained my relationship with the Church of God throughout high school, college, seminary, and graduate school, always seeking out a local Church of God congregation as a church home away from home. I was saddened by the criticisms and scorn of classmates who despised having been brought up in my church or in one like it. In retrospect, I can say that my early worship and leadership experience in the church equipped me for participation in the black student movement of the late 1960s and 1970s, especially those aspects of the movement that required planning, organization, public speaking, singing, or rehearsals. During my years at Swarthmore College, where my minor field of concentration was black studies, I began to observe the connections between slave religion and the contemporary Sanctified church. Throughout my years in seminary and graduate school, I remained skeptical of the liberal tendency to dismiss out of hand anything identifiable as conservative, evangelical, or fundamentalist, yet the liberal milieu of Harvard Divinity School gave me space for pursuing my own interests in Afro-American studies and religion.

My 1985 doctoral dissertation at Harvard was a study of the religious commitments and social ethics of the North American ex-slaves who were Christians. I carefully examined the conversion experiences and religious practices of these Christians and scrutinized the ways in which they envisioned their sufferings and observations as slaves in light of Christian testimony. Tom Ogletree's notion of preexilic and postexilic eschatology, as developed in *The Use of the Bible in Christian Ethics,* is cited in my concluding assessment of the ongoing social and ethical dilemmas faced by the aging ex-slaves during the first decades of the twentieth century as their oral histories were being collected and preserved. In a sense, the present study is a continuation of that inquiry, this time with an expanded view of the concept of exile and with attention focused on the late twentieth-century experiences of the physical and spiritual progeny of the generation of African American Christians who negotiated the transition from slavery to freedom.

The concept of exile as a category of meaning for African American people in the United States has been mentioned in many nineteenth- and twentieth-century interpretations of the African American experience. Yet, for African American Christians, the premier referent for the experience of exile is the Old Testament story of the forced deportation of the Jews into Babylon during the sixth century, B.C.E. Exilic images and ideas appropriated from the Psalms, the book of Job, and the prophetic writings of Isaiah, Jeremiah, Ezekiel, and Daniel have been fixed in the folk songs and sermons of the black churches. One of the key exilic texts is Psalm 137, which expresses the humiliation, anger, and despair of a captive people in poignant, poetic language. Reflecting on the religion, ethics, and culture of the Sanctified church tradition, the present investigation outlines an African American reading of the question raised by the Jews in Babylonian exile, "How could we sing the Lord's song in a foreign land?" (Psalm 137:4, NRSV).

In the pages that follow, my intention is to offer an analysis and description of seven dimensions of the African American experience of exile in successive chapters: social, religious, liturgical, cultural, institutional, intellectual, and ecclesial. Each of these dimensions is expressive of a pair of corresponding features or themes whose interplay characterizes that particular aspect of exilic existence in a dialectics of apparent opposites in creative tension. The introduction describes the emergence of the Sanctified church tradition as a twentieth-century African American Christian reform movement rooted in African and slave religious practices, whose definitive dialectics is the testimony of the saints as being "in the world, but not of it." The social dimension of exile is discussed in the first chapter, which sketches the early history of the Holiness and Pentecostal churches in terms of the dialectics of egalitarian and exclusivist notions of race, sex, and class. The second chapter profiles the ministry and worship of a contemporary Holiness congregation to portray the religious aspect of exilic existence in a dialectics of refuge and reconciliation. Liturgical exile is described in the third chapter in terms of a phenomenological analysis of the saints at worship as governed by a dialectics of static and ecstatic forms of spirit possession. The fourth chapter deals with competing sacred and secular trajectories in gospel music, a genre expressive of cultural exile. Institutional exile is the subject of the fifth chapter, which examines collegiate gospel choirs and black clergy caucuses in light of the dialectics of protest and cooperation. The problem of intellectual exile is addressed in the sixth chapter, with a focus on the dialectics of African and American identity as played out during the flowering of black consciousness and black power during the 1960s and 1970s. The book's seventh and final chapter

ventures a constructive exploration of the exilic ecclesiology of the Sanctified churches, toward the end of suggesting ways in which others can be
invited to embrace exilic identity as an ecumenical strategy for spiritual
empowerment, cultural survival, moral regeneration, and social transformation at the brink of the twenty-first century. The conclusion references
the African American intellectual tradition, gospel music, and contemporary biblical hermeneutics in a final reflection upon the overarching dialectics of exile and homecoming.

Washington, D. C.   C.J.S.
January 31, 1995

Acknowledgments

I did most of the research and writing for this book during my 1993–1994 sabbatical leave from Howard University, and I received financial assistance from two sources: the Theological Scholarship and Research Award from the Association of Theological Schools and a University-Sponsored Faculty Research Grant in the Humanities, Education, and Social Sciences from Howard. So many people at the School of Divinity have helped me along the way, including present and former deans, the administrative staff, the Divinity librarians and assistants, the personnel of the Information and Services Clearinghouse and Research Center on African American Religious Bodies, and my faculty colleagues and students.

I am grateful for my friend and teacher Cheryl Townsend Gilkes, whose encyclopedic knowledge of the Sanctified church tradition has been an inspiration. I thank my former dean Lawrence N. Jones and my dear friend Harold Dean Trulear for their careful reading of the entire manuscript. I have been sustained in this task by fond memories of the late Pearl Williams-Jones, whose scholarly and musical gifts still minister to me. Thanks to Evelyn Simpson Curenton and Mark A. Dennis.

My church family at Third Street Church of God has been an abiding source of support, information, and encouragement. I especially thank Lottie E. Giles and the Reverend Dr. Victor B. Phillips Sr. for assisting me with historical materials; Sandra D. Key, our minister of music and fine arts; and Vivian Owens Reid, who gave me my first opportunity as a teenager to sing gospel music in a choir. Pastor Samuel G. Hines inspired so much of what I have written; I deeply regret that he did not live to see this book come to print.

My parents, Doris and Wallace Sanders, must be thanked for all they have done for me from the beginning. I appreciate my brother, Eric, who made sure I knew how to read and count before I entered kindergarten. My husband, Alan Carswell, gave invaluable technical assistance and has been extremely tolerant of my preoccupation with this project and my incessant clutter of books and papers. My two children added "texture" to this study in their own unique manner. One day, as Allison rode her bicycle in circles around the room and Garrett used his version of the martial arts to fight imaginary aliens in the foyer, I can recall looking up from my portable computer at the dining room table and marveling at the possibility of actually completing this manuscript. There is a God.

Contents

Saints in Exile

Introduction:
"In the World, but Not of It"

One of the earliest uses of the term "Sanctified church" occurs in the anthropological writings of Zora Neale Hurston. Her book *The Sanctified Church* is mainly an anthology of assorted primary source materials compiled and published posthumously in 1981 that features only one brief essay bearing the same title. In that essay, she defines the Sanctified church as a revitalizing phenomenon that had arisen among various groups of "saints" in America. It is not a new religion and "is in fact the older forms of Negro religious expression asserting themselves against the new." Moreover, the Sanctified church is a "protest against the high-brow tendency in Negro Protestant congregations as the Negroes gain more education and wealth."[1] She identifies two branches of the Sanctified church: the Church of God in Christ and the Saints of God in Christ. The Sanctified church is closely related to three distinct Old and New World religious traditions: African religion, white "protest Protestantism," and Haitian *vaudou*. In this regard, Hurston sees "shouting" as nothing more than a continuation of the African possession by the gods, acknowledges the existence of "strong sympathy" between the white and Negro "saints," and notes the similarities between the dance of the saints and the steps seen in Haiti when a man or a woman is "mounted" by a *loa*, or spirit.[2] Sociologist Cheryl Townsend Gilkes has defined the Sanctified church as a segment of the black church that arose in the late nineteenth and early twentieth centuries, beginning at the end of Reconstruction, in response to and largely in conflict with postbellum changes in worship traditions within the black community. Its distinguishing mark is adherence to the traditions of oral music and ecstatic praise associated with slave religion. The label "Sanctified church" emerged within the black community to distinguish congregations of "the saints" from those of other black Christians, especially the black Baptists and Methodists who assimilated and imitated the cultural and organizational

3

models of European-American patriarchy. This label acknowledges the sense of ethnic kinship and consciousness underlying the black religious experience and "designates the part of the black religious experience to which a saint belongs without having to go through a dizzying maze of organizational histories involving at least twenty-five denominations."[3]

The picture is even more complex than Gilkes indicates; more than 100 church bodies listed in the *Directory of African American Religious Bodies* can be identified with the Sanctified church tradition, and a few of these actually include the word *Sanctified* in their official names, for example, Christ Holy Sanctified Church of America, Inc., Christ's Sanctified Holy Church, Church of God (Sanctified Church), and Original Church of God (or Sanctified Church).[4] However, the *Directory* does not use the term "Sanctified" as a general category and, instead, employs the more cumbersome and perhaps more technically accurate designation "Pentecostal/Apostolic, Holiness, and Deliverance." The article presented as a historical overview of this category of churches in the *Directory* was written by theologian William C. Turner Jr., who emphasizes the reform aspects of the Holiness/Pentecostal/Apostolic churches as a "segment of the black church that desperately sought restoration and revival through a recovery of latent spirituality."[5] These churches were led by black Christians around the turn of the century who "came out" of the mainline black denominational churches and sought "the deeper life of entire sanctification" and Spirit baptism: "Their initial concern was not so much to start a new denomination as to call the existing ones back to the wells of their spirituality."[6]

Turner cites Leonard Lovett's list of the five original groups in black Holiness-Pentecostalism, shown here in order of date founded: (1) United Holy Church of America (1886), (2) Fire Baptized Holiness Church of God in the Americas (1889), (3) Church of Christ Holiness, U.S.A. (1894–1896), (4) Church of God in Christ (1895–1897), and (5) Pentecostal Assemblies of the World (1914–1924).[7] Turner clearly delineates the similarities and differences of the various groups within the Holiness/Pentecostal/Apostolic movement. A historical chronology is implicit in the categories. Four of the five churches—the United Holy Church of America; the Church of Christ Holiness, U.S.A.; the Church of God in Christ; and the Fire Baptized Holiness Church of God in the Americas—were originally Holiness bodies formed before the turn of the century around the doctrine of sanctification. Three of these—the United Holy Church of America, the Church of God in Christ, and the Fire Baptized Holiness Church of God—embraced Pentecostal teachings and practices after the 1906 Azusa Street Revival in Los Angeles, following the lead of William J. Seymour. The Church of Christ Holiness, U.S.A. remained a Holiness group under the leadership of Charles Price Jones after he parted company with Charles Harrison Mason, founder of the Church of God in Christ. The Pentecostal Assemblies of the World had its beginnings as a product of the Azusa Street Revival—which is to say, as a Pentecostal body—but took on the Apostolic doctrine and identity in 1914 after the rebaptism of G. T. Haywood of Indianapolis and his followers "in Jesus' name."

What the Holiness, Pentecostal, and Apostolic churches all have in common is an emphasis on the experience of Spirit baptism. A crucial point of disagreement is whether a person must speak in tongues (glossolalia) to validate his or her Spirit baptism; "the tongues doctrine unites Pentecostals and Apostolics, and divides Holiness believers from the former two."[8] Some Holiness believers reject glossolalia altogether; others appreciate and/or practice speaking in tongues without insisting on the doctrine of tongues. The Holiness emphasis is on sanctification, or personal holiness, whereas the Pentecostals and Apostolics emphasize spiritual power. Apostolics differ from the other two groups primarily on theological grounds. They reject the doctrine of the Trinity, adhering instead to belief in the "oneness of God" as revealed in Jesus Christ. They do not use the trinitarian formula (i.e., "in the name of the Father, the Son, and the Holy Ghost") for baptism as is characteristic of other Christian churches; they baptize in Jesus' name. Hence, the Apostolics are occasionally referred to as "Jesus only" or "Jesus' name" churches. Although some of these churches practice speaking in tongues and some baptize only in Jesus' name, all adhere to some form of doctrine and practice of sanctification; thus, the term "Sanctified church" is inclusive of them all. Historically, these churches have been known to preach and promote an ascetic ethic forbidding the use of alcohol, tobacco, and other addictive substances, gambling, secular dancing, and the wearing of immodest apparel.

In view of this understanding of the Holiness, Pentecostal, and Apostolic churches as components of a historical movement among African American Christians to promote sanctification and Spirit baptism in some distinct form, it is now possible to formulate a comprehensive definition of the Sanctified church that builds on the thought of Turner, Hurston, and Gilkes but adds a needed ethical dimension: *The Sanctified church is an African American Christian reform movement that seeks to bring its standards of worship, personal morality, and social concern into conformity with a biblical hermeneutic of holiness and spiritual empowerment.* This ethical emphasis is a critical element in the definition because the Sanctified churches are congregations of "saints," an ethical designation members apply to themselves as an indication of their collective response to the biblical call to holiness. The saints follow the holiness mandate in worship, in personal morality, and in society, based on a dialectical identity characteristic of the tradition: "in the world, but not of it." This dialectical identity reflects the social aspect of exilic consciousness, as manifested in the saints' awareness of alienation or separation from the dominant culture, based on racial differences and religious practices.

African Religious Traditions in the Sanctified Church

As Hurston observed, the practices of shouting and spirit possession in the Sanctified church movement are readily associated with African traditional religions and diasporic worship practices. Afrocentric philosopher Molefi Asante, for whom Pentecostal worship was a formative influence, stated in his book *Afrocentricity* that "the music and dance of the church may be the essence of our Africanity."[9] Without specifically naming the Pentecostal or Sanctified tradition, Asante describes the "general experience of those traditional black churches which have emerged out of the roots of our past" as "truly an African expression":

> The panorama of Africa is not merely unfolded but expanded and amplified in the religious drama. More than this, the church services become a collective outpouring of the soul with some people getting more possessed than others but no one really escaping the influence of possession even if it is no more than the slight tapping of the foot. Syncopated pianos and organs and hand-clapping often drive the faithful into ecstasy. The rhythms run to be free, individuals shout and moan, the preacher directs this "mass madness," which is really not madness, by the call and response, and suddenly the whole congregation is praising the Lord.[10]

He claims that the African slaves, whose manner of "getting religion" was imitated by white evangelicals, were actually replicating in America the same ecstatic religious behavior practiced in Africa. What the Africans were "getting" was the same "ecstatic combinations produced by the polymeters of African music. In the place of drums the African-American substituted hand-clapping, foot-stomping, head-shaking, body-moving rhythm—all in an attempt to drive the self into further possession, by the Lord."[11] Asante offers this portrayal of the distinctive worship practices of the Sanctified church tradition as his best evidence to support an essentialist argument for the unique nature of the African people, at home or in diasporic communities. He seems less interested in the ethical meaning of these worship practices in the American context, except to note that they were imitated and not initiated by whites.

In his 1978 doctoral dissertation comparing black political and religious movements, Pentecostal political scientist James S. Tinney identifies three dominant Africanist themes with black Pentecostalism: spirits, magic, and eschatology. In this cosmology, the functions of good spirits, or angels, are absorbed by one divine Spirit, the Holy Ghost. The functions of the supremely evil spirit known as "the Devil" or "Satan" are assumed by legions of evil spirits called "demons." Thus, an eternal conflict is presupposed between demons and the Holy Ghost:

> Worship services may often begin with an exorcism ("cast out the demons of fear, cast out the demons of despair, cast out the demons of unbelief"). Illnesses, physical or psychological compulsions, and "besetting sins" are also subject to exorcisms. But so is the environment, the social and political order. Politics is not evil, but it is subject to control by the Holy Ghost or demons. The same goes for political communities. The kingdom of God is nothing more or less than the world in which the Holy Ghost reigns.[12]

Although Tinney acknowledges that the term "magic" is not used in these churches, he nevertheless argues that Pentecostal rituals for supernatural healing directly repeat African practices in an effort to "tap and harness spiritual power" in order to make it serve the believer in a beneficial way. It should be noted that most of the healing practices Tinney designates as African and magical—such as prayer, anointing the sick with oil, laying on of hands, and the commission of special cloths or handkerchiefs—are also found in the New Testament.[13] The African eschatology Tinney ascribes to black Pentecostals is a realized eschatology that claims in this life those things that religions generally hold as rewards for the faithful in the afterlife.

> One does not wait to die in order to attain perfection; he can attain it in this life. One does not wait for death to bring relief from illness and disease; healing and miracles are offered now. One does not wait to receive spirit energy in another world; it, too, is available in the Holy Ghost experience now.[14]

In a separate monograph, Tinney offers further corroboration of the relationship between black Pentecostalism and traditional African religions by describing the receptivity of Africans to the Pentecostal revival: "When news of the [Pentecostal] revival reached Africa, it was not regarded as an entirely 'new' phenomenon. For the very practices of Azusa Street had long been known in the Motherland even before any Azusa-inspired messenger arrived."[15] He has in mind here several African independent Christian groups of the "Pentecostal variety," such as the movement led by Prophet Harris, which had already emerged without white supervision in West Africa prior to the Pentecostal revival. He estimates that the majority of the more than 6,000 independent church organizations in Africa are Pentecostal. Ethically, these churches bear similarity to the Sanctified church tradition in America with respect to their cultural and political resistance to white imperialists and missionaries.

John Philips has commented that few historians, ethnographers, or sociologists of religion have explored the African origins of white religious belief and behavior. He is of the opinion that it is no easy matter to distinguish the uniquely African elements of Pentecostal and Holiness worship practices because much that is often considered peculiarly black about black Pentecostal and Holiness churches is often equally characteristic of white churches. The characteristic features of Pentecostal churches that are demonstrably African in origin include possession trances, ritual dancing, drumming, and ecstatic speech (thought to be the language of angels or spirits).[16] Philips asserts that, in order to write the definitive statement about African cultural survivals in the United States, one must acquire a triple expertise—as an Americanist, an Africanist, and a Europeanist.[17] His conclusion that "as much African culture survives now among whites as among blacks" pushes him to endorse anthropologist Melville Herskovits's view that much African culture survives in the United States and at the same time to assent to sociologist E. Franklin Frazier's argument that African culture is not the distinguishing characteristic of African-American society.[18]

Adopting a "hemispheric perspective," historian of religion Joseph Murphy sees black religion in the United States as a special articulation of an African-derived spirituality that has kindred expressions throughout the Americas.[19] Murphy compares five traditions—Haitian *vaudou,* Brazilian *candomblé,* Cuban *santería,* Revival Zion in Jamaica, and the black church in the United States—in somewhat ahistorical broad strokes. He uses a Church of God in Christ congregation in Washington, D.C., as his primary source of data representing the black church, a choice based on the impression that the Pentecostals alone have preserved the shout and spirit possession in the United States. Murphy makes the important assumption that what might seem to outsiders to be mere "motor behaviors" remembered from Africa are actually expressions of a dynamic, incarnated spirituality found throughout the diaspora.[20] He is aware of the relative weakness of surviving African religious customs in the United States compared with those in the Caribbean and South America, a distinction that also rests at the root of the classic debate between Herskovits and Frazier.

The emphasis on spirituality rather than theology in the comparative study of diasporic religious traditions enables Murphy to make an interesting point concerning differences in denotation of racial terms as applied to the church: "The Black Church is both 'black' in its independent wisdom arising from its exclusion from white America, and it is 'African American' in its development of a spirituality born in Africa."[21] This line of thought encourages reflection on the ethical meaning of the Sanctified church tradition. Murphy believes that the religious traditions of the African diaspora are alike in that each shares a social history of enslavement and racial discrimination, and each became the focus for an extraordinary struggle for survival against and triumph over brutal systems of exploitation. Moreover, they share an elevated sense of solidarity against injustice and a commitment to the protection and advancement of their communities, understood as the "work" of the "spirit."[22] Importantly, he proffers this ethical evaluation of the roles these traditions have played in fostering cultural and political change without relinquishing the belief that "the spirit is a real and irreducible force uplifting communities throughout the African diaspora."[23]

Christianity and Social Ethics among the Slaves

It seems that the religion of the African Americans in bondage should not be designated as slave religion, any more than Judaism should be called slave religion because of the formative role a slave community played in its early history. The religion of the enslaved Christians was an amalgam of Christian and African practices and beliefs, with the slaves' distinctive African practices noted more frequently in the sources than their generic Christian beliefs. However, the slaves who were Christian were very forthright about identifying themselves as Christians, as can be readily observed in published collections and analyses of ex-slave narratives and oral histories, such as *Cut Loose Your Stammering Tongue, Deep Like the Rivers, God Struck Me Dead,* and *Weevils in the Wheat.*[24] The Christian community in bondage produced a significant social and ethical critique of the hypocrisy of white Christianity, insofar as it promoted and justified slaveholding. This critique was thoroughly documented by the painful scars of their own flesh and by the stories and memories of collective suffering at the hands of people who called themselves Christians. It is important to appreciate that there were Christians within the slave community who saw themselves as having a special social-ethical imperative: to seek freedom by resisting, in the name of Christ, the social, cultural, ecclesial, and philosophical structures whites had erected, in the name of Christ, to keep them in bondage.

One noteworthy example of the tradition of resistance among enslaved Christians is that of Harriet Tubman. The fact that she led hundreds of slaves in the escape from bondage to freedom is widely known. What is hardly ever addressed in the many references to her heroism is the depth of her Christian faith and conviction. In a fit of rage, Tubman's slave master had thrown a heavy weight at her, breaking her skull and inflicting severe brain damage that plagued her from childhood to old age with a sleeping disorder. At one point she became bedridden for several months, and in this condition she prayed unceasingly for her master to be converted to Christianity. Her prayers of intercession took an interesting turn when she learned of her master's plans to sell her:

> Den I heard dat as soon as I was able to move I was to be sent with my brudders, in the chain-gang to de far South. Then I changed my prayer, and I said, "Lord, if you ain't never going to change dat man's heart, *kill him,* Lord, and take him out of de way, so he won't do no more mischief." Next ting I heard ole master was dead; and he died just as he had lived, a wicked, bad man.[25]

She regretted that her prayers were responsible for her master's death, but at that point she began to recover and to seek sanctification:

> 'Pears like, I prayed all de time, about my work, eberywhere; I was always talking to de Lord. . . . When I took up de broom and began to sweep, I groaned, "Oh, Lord, whatsoebber sin dere be in my heart, sweep it out, Lord, clar and clean; but I can't pray no more for pore old master.[26]

Her prayers for purification from sin thus superseded her prayers of intercession for her master.

Tubman began to see visions of horsemen and shrieking women and children being torn from each other and "a line dividing the land of slavery from the land of freedom, and on the other side of that line she saw lovely white ladies waiting to welcome her, and to care for her."[27] Guided only by the North Star, she set out to find liberty. Her ethical rationale for seeking freedom and for using violence, if necessary, to secure it was clearly reasoned out and articulated in terms of doing the will of God:

> I had reasoned dis out in my mind; there was one of two things I had a *right* to, liberty, or death; if I could not have one, I would have de oder; for no man should take me alive; I should fight for my liberty as long as my strength lasted, and when de time came for me to go, de Lord would let dem take me.[28]

When Tubman succeeded in making her escape, she crossed the line in fulfillment of one aspect of her vision of freedom, but there were no "lovely white ladies" to welcome her. She used the language of exile to describe her initial experience and impressions of freedom, again with prayer as her principal context for both thought and action:

> I had crossed de line of which I had so long been dreaming. I was free; but dere was no one to welcome me to de land of freedom, I was a stranger in a strange land, and my home after all was down in de old cabin quarter, wid de ole folks, and my brudders and sisters. But to dis solemn resolution I came; I was free, and dey should be free also; I would make a home for dem in de North, and de Lord helping me, I would bring dem all dere. Oh, how I prayed den, lying all alone on de cold, damp ground; "Oh, dear Lord," I said, "I haint got no friend but *you*. Come to my help, Lord, for I'm in trouble."[29]

The task of liberating others from bondage can be seen as Tubman's constructive response to the experience of exile. She solicited God's help in her efforts to create a community of exiles who could welcome each other to the land of freedom. There is a clear correlation between Tubman's stages of spiritual development and her stages of preparation for the work of liberation through escape and escort, between the intensification of her life of prayer and her deepening resolve to liberate her people. Her work was carried out with a keen sense of utter dependency on God, and the testimony of sanctification emerged as a significant element of her spiritual and political formation.

The single best account of the Christian faith and collective spiritual experiences of the slave community in the United States remains Albert J. Raboteau's *Slave Religion* (1978). Although Raboteau does not mention the Sanctified tradition, he is aware of the ethical dilemma faced by slaves who wanted to dance their religion in conflict with the mores of white and black evangelicals:

> Despite the prohibition of dancing as heathenish and sinful, the slaves were able to reinterpret and "sanctify" their African traditional dance in the "shout." While the North American slaves danced under the impulse of the Spirit of a "new" God, they danced in ways their fathers in Africa would have recognized.[30]

Raboteau directs his attention to the shout through the critical eye of African Methodist Episcopal Bishop Daniel Payne, author of what has become the classic statement of the black evangelical rejection of black folk religion and worship styles:

> About this time [1878] I attended a "bush meeting.". . . After the sermon they formed a ring, and with coats off sung, clapped their hands and stamped their feet in a most ridiculous and heathenish way. I requested the pastor to go and stop their dancing. At his request they stopped their dancing and

> clapping of hands, but remained singing and rocking their bodies to and fro.
> This they did for about fifteen minutes. I then went, and taking their leader
> by the arm requested him to desist and to sit down and sing in a rational
> manner. I told him also that it was a heathenish way to worship and disgrace-
> ful to themselves, the race, and the Christian name.[31]

Raboteau suggests that the holy dance of the shout may have been a "two-
way bridge connecting the core of West African religions—possession by
the gods—to the core of evangelical Protestantism—the experience of
conversion." In the book he divides his attention between these two "core"
elements in slave religion and describes with care the impact of both the
African religious heritage and American evangelization efforts by using a
host of primary and secondary sources representing the vantage point of the
slave. Yet, he concludes that, notwithstanding continuities in performance
styles from Africa to America among the slaves, "in the United States the
gods of Africa died."[32]

    In Raboteau's view, the influence of African religion on the Protestant
revivalist tradition is more a matter of style than content. He draws a similar
conclusion concerning the development of the Negro spirituals: "African
style and European hymnody met and became in the spiritual a new, Afro-
American song to express the joys and sorrows of the religion which
the slaves had made their own."[33] After presenting a strong case for
the distinctive nature of slave religion, with an emphasis on the slaves'
ethical sensitivity to white Christian inhumanity, brutality, and hypocrisy,
Raboteau somewhat forces the conclusion that there was a genuine religious mutuality between black and white Christians during slavery. His
description of this religious mutuality reflects the practices and character of
the holiness tradition:

> Religion, especially the revivalistic, inward, experientially oriented religion
> to which many slaves and masters adhered had an egalitarian tendency which
> occasionally led to moments of genuine religious mutuality, whereby blacks
> and whites preached to, prayed for, and converted each other in situations
> where the status of master and slave was, at least for the moment, suspended.
> In the fervor of religious worship, master and slave, white and black, could
> be found sharing a common event, professing a common faith and experienc-
> ing a common ecstasy.[34]

Curiously absent from this discussion of mutuality is evidence of white
Christian repentance for sins perpetrated against the slave.
    Writing against the view that slave religion was merely compensatory
otherworldliness, Raboteau evaluates the distinctive moral ideas of the
slave community and the attitude of moral superiority some of them had in

relation to whites. Fundamental to the slave community's ethics is the rejection of the master's religion as hypocritical, which becomes a basis for some of them to refuse to obey moral precepts held up to them by whites and to justify stealing, lying, extramarital sexual relations, and other violations of "white" Christian mores. Raboteau contrasts the moral behavior of the sinners and saints among the slaves with some ambivalence. Although some slaves rejected the moral system preached by the master and his preachers, others devoted themselves to a life of virtue in which they developed both a sense of personal dignity and an attitude of moral superiority to their masters, a morally suspect attitude that could simultaneously support compliance to the system of slavery and buttress the slave's own self-esteem.[35] Raboteau describes the actions and attitude of a pious Christian slave who forgave his master for wrongly punishing him for something he had not done, thereby acquiring "the leverage of moral virtue by which to elevate his own self-worth," and who subsequently escaped slavery with a somewhat guilty conscience.[36] Some pious slaves drew a distinction between sacred and secular dancing but would be forced to dance against their will to please their masters during the Christmas holiday celebrations.[37] Moreover, prayer became an important symbol of resistance for a people forbidden to pray. There are instances in which slaves were whipped and punished for praying. During the Civil War, some whites even attempted to coerce their slaves into praying for the success of the Confederacy.[38]

Sterling Stuckey's *Slave Culture* lifts up one specific aspect of slave religion—the ritual of the shout—in a general discussion of nascent black nationalism. He attempts to make a connection between the African nationalism of the slave community, consisting of values that bound slaves together and sustained them under brutal conditions of oppression, and the nationalist theory developed by David Walker, Henry Highland Garnet, W. E. B. Du Bois, and Paul Robeson. Stuckey sees the ring shout as a ritual of black unity that not only enabled the slaves to overcome barriers of language and ethnic difference but also in the twentieth century "continued to form the principal context in which black creativity occurred."[39] He celebrates the importance of the ring shout for jazz musicians like Milt Jackson and Thelonious Monk, who were familiar with it because they lived near one of the Sanctified churches during childhood. Yet, his only example of its importance in contemporary sacred culture is taken from James Baldwin's novel *Go Tell It on the Mountain*. Stuckey says of the black nationalist theorists, "Most were exposed to main currents of African culture without understanding how those currents might contribute to the surge toward liberation they wanted to initiate."[40] It seems that the key to understanding these currents and their liberating potential is to recognize, first of all, the religious and ethical meanings that cultural practices like the ring shout have for the people who are engaged in them. Moreover, that shouting is praise may be totally incomprehensible to people who never practice it, and this simple fact must be kept in mind. Stuckey takes great care to show how much exposure Walker, Garnet, and the others had to ring shouts during their childhood or at some other time. However, he does not seem to appreciate the necessity of maintaining a sacred interpretation of this ritual in addition to observing its secular implications and effects.

Harriet Tubman provides an illuminating example of the discrepancy between the religious sensibilities of the cultural nationalist and the religious "folk."

> She rose singing, *"My people are free!" "My people are free!"* She came down to breakfast singing the words in a sort of ecstasy. She could not eat. The dream or vision filled her whole soul, and physical needs were forgotten. Mr. Garnet said to her: "Oh, Harriet! Harriet! You've come to torment us before the time; do cease this noise! My grandchildren may see the day of the emancipation of our people, but you and I will never see it." "I tell you, sir, you'll see it, and you'll see it soon. My people are free! My people are free."[41]

Obviously, Garnet did not appreciate Tubman's ecstatic, exuberant display of emotion at the breakfast table, he did not comprehend the reason why she rejoiced, and he was not convinced that her vision of freedom had any authority or veracity. From her perspective, however, the manner, rationale, and content of her praise were entirely in order because she was overwhelmed by the belief God had given her a revelation of the emancipation of her people. Nationalist theorists and historians alike will continue to underestimate the depths of African culture in America, as Stuckey claims they have done, as long as they persist in writing off the religious meaning that shouting and other black folk rituals have in the minds of the believers, while basing their theories solely on the grounds of the social, cultural, and political significance these practices have in the minds of unbelieving intellectuals.

Emergence of the Sanctified Church
as a Christian Renewal Movement

David Daniels, a church historian who is an ordained minister in the Church of God in Christ, has written a doctoral dissertation on the Holiness movement in Mississippi as a renewal movement among African American

Baptists from 1895 to 1905. He is concerned to show how the Holiness movement negotiated the deconstruction of the theological world of slave religion and the reconstruction of a new theological world, making it a third force in African American Christianity.[42] Daniels posits his "third force" argument over against Gayraud Wilmore, James Tinney, and Cheryl Townsend Gilkes, who, in his view, fail to see the Holiness movement as a reform movement because they have adopted Carter G. Woodson's "two-force" scheme and identify the movement with the conservatives or traditionalists (bound by ignorance, dogmatism, otherworldliness, and emotionalism) and not with the progressives (committed to education, innovation, and this-worldliness): "It is more than the perpetuation of slave religion as Wilmore notes, or a mere expression of slave religion as Tinney and Lovett argue, or a reaction to the progressive reforms being undertaken in African American Christianity as Townsend-Gilkes contends."[43] Daniels argues that the Holiness movement in Mississippi, under the leadership of Charles Price Jones and Charles Harrison Mason, was able to reject the negative attitude toward slave religion that the progressives held and advocate the moral, ecclesial, liturgical, and pastoral reforms that the progressives embraced. One of Jones's principal contributions to the movement was linking salvation to the formation of character, which he envisioned as a more important foundation of society than family, politics, economics, or even the church.[44] Bishop Mason, founder of the Church of God in Christ, defended a continued utilization of the liturgical practices of slave religion, in particular, "the elements of the rituals associated with the ring-shout, supplemented by moral reforms and the experience of sanctification."[45]

A similar argument is made with respect to the Azusa Street Revival: It was a product of former slave William J. Seymour's restructuring of slave religion and not just a product of slave religion generally defined. This restructuring argument is significant because it goes beyond the observations made by Raboteau, Stuckey, and others that African worship styles and practices were preserved in slave religion and in subsequent developments in black culture. Daniels gives important attention to the substance and content of black folk religion and acknowledges critical changes and transformations black Christians effected in their religion as they made the transition from slavery to freedom. Unlike most other scholars, who have analyzed the ring shout in terms of emotional release or African-derived spirit possession, he gives priority to the religious meaning of this practice and to the ways in which it has been adapted into contemporary Sanctified church rituals, such as tarrying (a ritual invocation of the manifestations of the Holy Spirit) and the holy dance.

At the turn of the century, the ecclesiastical courts in the local Baptist associations and state convention of Mississippi, upheld by the state Supreme Court, disfellowshiped Jones and Mason because their teachings concerning the nature of the church were no longer palatable to their Baptist clergy brethren. By this process, the Holiness movement was expelled from the Baptist denomination and, in effect, became marginalized in African American Christianity.[46] Once Jones, Mason, and their followers were expelled by the Baptist denominational authorities, they gained new authority and insight to engage in the reformation of American Christianity and the advancement of African American people. In other words, they were able to make themselves at home in the Sanctified church, a sacred space "in the world, but not of the world," where meaningful worship traditions could be preserved, practiced, and produced unencumbered by charges of heathenism or heresy.

1

The Sanctified Churches and Christian Reform: Confronting the Barriers of Race, Sex, and Class

One of the distinctive contributions the Sanctified church movement has made to the historical evolution of North American Protestantism is the involvement of blacks, women, and the poor at all levels of its ministry. Although some of the mainstream churches are beginning to ordain and appoint women pastors in significant numbers and are showing some interest in multiculturalism and outreach to the poor, as the twentieth century draws to a close, the Sanctified church begs for recognition as a place where poor black women and men have been empowered to do ministry since its inception during the late nineteenth century. The shifting patterns of inclusion and exclusion in these churches are governed by two primary factors, namely, the egalitarian doctrine of the Holy Spirit, on the one hand, and the impact of racist, sexist, and elitist societal norms on the other.

Configurations of Race and Denomination in Holiness, Pentecostal, and Apostolic Churches

Considered as a matter of ethical concern, the dynamics of race in the Holiness, Pentecostal, and Apostolic churches may have more to do with the orientation toward interracial ministry than with the actual levels of white and black participation. The five "original" bodies that constitute the Sanctified church tradition—the United Holy Church of America, the Church of Christ Holiness, U.S.A., the Church of God in Christ, the Fire Baptized Holiness Church of God in the Americas, and the Pentecostal Assemblies of the World—share a similar history with regard to race.[1] William C. Turner Jr. has made the claim that the typical pattern of emergence of a Sanctified church body is that an individual or a group of African American Baptists or Methodists involved themselves in some sort of prayer meeting or union service to recover the intensity and fervor of their spiritual life, especially seeking "the deeper life of sanctification."[2]

From the time when these black Holiness and Pentecostal bodies were first formed until the present, the vast majority of black Christians have been Baptist or Methodist. Several factors attracted blacks to the Baptist and Methodist churches, which represent, as Cornel West has said, the two major traditions of dissenting American Protestantism. The Baptist churches provided many black slaves with a sense of somebodiness, a personal and egalitarian God who gave them an identity and dignity not found in American society, and control over their own ecclesiastical institutions: "The uncomplicated requirements for membership, open and easy access to the clergy and congregation-centered mode of church governance set the cultural context for the flowering of Africanisms, invaluable fellowship, and political discourse."[3] The Baptist doctrinal "heritage" of the African Americans who eventually formed the Sanctified church tradition includes the notion of the priesthood of all believers, which allowed that "the Holy Spirit testifies directly and immediately to the believer's heart, bypassing priest and teacher." When the autonomy and free expression of the Baptists were joined with the Methodist emphasis on holiness, what eventually emerged was African American Pentecostalism.[4]

As attractive as these Baptist and Methodist teachings may have been, separate black Baptist and Methodist religious bodies came into existence as a consequence of black Christians being subjected to racist and exclusivist practices in the white Baptist and Methodist churches, beginning in the eighteenth century. The first separate black Baptist church was organized between 1773 and 1775 in Silver Bluff, South Carolina, and its nucleus included David George, a slave who preached there regularly.[5] Blacks have been involved in Methodism from its inception in the American colonies prior to the Revolutionary War.[6] Richard Allen and Absalom Jones led the black Methodists out of St. George's Methodist Episcopal Church in Philadelphia in 1787 after being dragged away from the altar while at prayer.[7] The six largest mainstream black denominations—African Methodist Episcopal; African Methodist Episcopal Zion; Christian Methodist Episcopal; National Baptist Convention, U.S.A., Inc.; National Baptist Convention of America, Unincorporated; and Progressive National Baptist Convention—have all been partakers of the history of religious exile in America on the basis of denominational racism. "Denominational racism" here refers to the marginalization and/or segregation of adherents on the basis of race within a Christian denomination, with the sanction of its leaders and constituents.

One way of conceptualizing the emergence of the Sanctified church movement in light of the evolution of denominational racism in America is to observe two stages of black Christian alienation and response: first, blacks "came out" of the Baptist and Methodist churches because of racism; then, they "came out" of the black Baptist and Methodist denominations because of a commitment to holiness. Ironically, as they recovered, reformulated, and advanced the doctrine of sanctification and other teachings from the Methodist and Baptist traditions, the black saints were in some cases led back into fellowship with whites who had "come out" of the Protestant denominations. The emergent Holiness and Pentecostal groups, including the five original churches that engendered the Sanctified church movement, eventually fell into the same pattern, so that by 1924 most of them had become as rigidly segregated by race as the Baptist, Methodist, and other American Protestant churches.

Viewed from the historical perspective of denominational racism, the behavioral patterns of black and white Christians manifest distinctive social ethics. What is different in this history is that it is the whites who took the initiative to "come out" of the Holiness and Pentecostal churches that practiced interracial fellowship and ministry, for racial reasons rather than doctrinal ones. Further, it is evident that black leaders generally did not mimic the racist behavior of whites by alienating whites. White members left on their own accord, and the new denominations they started became, in time, more definitely "mainstreamed" among white American Protestants and racially exclusive, while the black groups maintained their opposition to white racism.

A brief background sketch of each of the five original denominations should suffice to illustrate five different patterns of denominational differentiation based upon issues of race and doctrine.

The first type is a black Holiness denomination that retains its black Holiness identity after the Pentecostal revival. The Church of Christ Holiness, U.S.A., was founded in 1894 by Charles Price Jones and Charles Harrison Mason, two ordained Baptist ministers who were disfellowshiped by the Baptists in Mississippi because of doctrinal differences. Jones continued to lead this body after he and Mason split subsequent to Mason's participation in the Azusa Street Revival and his adoption of Pentecostal teachings and practices.

A second type is a black Holiness denomination becoming a black Pentecostal denomination, such as the Church of God in Christ. When Mason and Jones parted company in 1907 over the doctrine of speaking in tongues, the Church of God in Christ was reorganized as a Pentecostal denomination in Memphis, Tennessee. Bishop Mason ordained scores of white ministers, and, in the years 1909 to 1914, there were as many white Churches of God in Christ as there were black, all bearing Mason's credentials and incorporation. Mason gave his blessing to the white ministers who formed the Assemblies of God in Hot Springs, Arkansas, in 1914, and maintained fellowship with white Pentecostals despite the racial separation.[8] Also in this category is the United Holy Church of America, which traces its beginnings to a black Holiness group convened in 1886 as the result of a revival conducted by Isaac Chesier in Method, North Carolina. White Pentecostals sought to direct the United Holy Church of America; in order to repudiate attempts at white control, a formal government commanded only by blacks was developed.[9]

A third type is an interracial Holiness denomination forming racially separate Pentecostal denominations. The Fire Baptized Holiness Church of God in the Americas shares a common heritage with the Fire-Baptized Holiness Church founded by Benjamin Hardin Irwin in Lincoln, Nebraska, in 1895. From 1898 to 1908, the church was interracial. After the Azusa Street Revival, the church amended its Holiness doctrine to include the Pentecostal view of tongues and thus became the first official Pentecostal denomination in the United States. Also in 1908, the blacks formed a separate body under the leadership of William E. Fuller, who was a convert from Methodism.

A fourth type is an interracial Pentecostal denomination forming racially separate Pentecostal denominations. The Pentecostal Assemblies of the World was formed in 1907, following the Azusa Street Revival, and was a fully integrated body until racial tensions with white members led to a schism in 1924. The whites split off under other names, and black leader Garfield Thomas Haywood of Indianapolis, Indiana, who had a Baptist and Methodist background, established the church's episcopal polity as the first bishop. In 1931, the Pentecostal Assemblies of Jesus Christ was formed in an effort to return to the original interracial fellowship of Oneness Pentecostalism, but by 1945 the remaining whites in the group merged with the Pentecostal Church, Incorporated, to form the United Pentecostal Church, International, and the blacks returned to the Pentecostal Assemblies of the World.[10]

A fifth type is an interracial Holiness or Pentecostal denomination with local congregations, caucuses, conventions, and/or educational institutions serving its black constituents. In these cases, black leaders and congregations within the white Holiness and Pentecostal bodies have maintained fellowship, credentials, and other official ecclesial ties with whites. Thus, there are black congregations and conventions within the Holiness and Pentecostal churches that exhibit most, if not all, characteristics of the Sanctified church tradition but are not identified with black denominations.

The *Directory of African American Religious Bodies* has entries for several of these interracial religious bodies or agencies, including the Assemblies of the Lord Jesus Christ, Inc.; the Church of God (Cleveland, Tennessee); the Church of God (U.S.A.); and the National Association of the Church of God (Anderson, Indiana).[11] The Church of God (Anderson, Indiana), merits special attention in this discussion for two reasons: its historical struggle to maintain interracial communion and its link to the 1906 Pentecostal revival through William J. Seymour.

A point of clarification is warranted here with respect to nomenclature. Because "Church of God" has been so frequently adopted from Acts 20:28 ("feed the church of God, which he hath purchased with his own blood") as a denominational name by Holiness and Pentecostal church groups, the conventional designation is to append to the title Church of God the location of the agency headquarters in parentheses or to use specific modifiers assigned by the group. It should never be assumed that similar titles imply similar identity. These conventions make it possible to distinguish between the various groups; for example, Church of God (Cleveland, Tennessee) is distinct from Church of God (Anderson, Indiana), and Church of God in Christ, Inc., is not the same as Church of God in Christ, International. Some of these church names are borrowed from a combination of Acts 20:28 and 1 Timothy 3:15–16a ("that thou mayest know how thou oughtest to behave thyself in the house of God, which is the church of the living God, the pillar and ground of the truth. And without controversy great is the mystery of godliness") and can be quite lengthy, but it is nevertheless important to acknowledge the distinctions among them. These titles and distinctions can get very complicated. For example, the Church of the Living God, the Pillar and Ground of the Truth, Which He Purchased with His Own Blood, Inc. (Lewis Dominion) is a group founded by Bishop Mary Magdalena Lewis Tate in 1903, and the House of God Which Is the Church of the Living God, the Pillar and Ground of the Truth without Controversy (Keith Dominion) was established in 1931 as a separate dominion or ecclesial body related to Bishop Tate. However, the House of God Which Is the Church of the Living God, the Pillar and Ground of Truth, has a totally different parent body, the Church of the Living God (Christian Workers for Fellowship) founded in 1889 by the Reverend William Christian.

"Zion's Hill": Black Holiness in the Church of God

Although the Church of God (Anderson, Indiana) in the United States is relatively small, with 221,346 members in 2,295 congregations as of 1995, the observation that as many as 20 percent of its congregations and 15 percent of its membership are black sets it apart from the dominant American pattern of denominational racism.[12] Moreover, in recent years, black ministers have occupied several major leadership positions within the group, including president of the chief governing body of ministers, chief executive officer, and dean of the graduate theological seminary at Anderson. Almost all of the black ministers and congregations are affiliates of an organization known as the National Association of the Church of God, which is not a separate denomination and does not grant ministerial credentials but does have its own administrative structures, physical plant, conventions, and educational ministries. Significant numbers of blacks attend the annual international convention of the Church of God each June in Anderson, as well as the quadrennial world conference of the Church of God convened in international sites as diverse as Europe, Asia, Africa, and Australia, but more blacks attend the two annual events sponsored by the National Association: the August camp meeting at West Middlesex, Pennsylvania, and the December youth conventions housed in hotels and convention centers throughout the United States. In terms of preaching, teaching, music, and worship style, these national gatherings are very much reflective of the Sanctified church tradition. The same can be said for Sunday worship services at numerous local Church of God congregations with a majority of blacks in attendance.

As early as 1917, at about the same time as other Holiness and Pentecostal bodies were resolving the race dilemma by splitting along racial lines, blacks in the Church of God established their own annual camp meeting in West Middlesex, Pennsylvania, near the Ohio state line. The core group that started this camp meeting association included Elisha and Priscilla Wimbish, a black Baptist couple from Cleveland. In 1904, Brother Wimbish had a vision of a camp meeting site, as told by his wife, Priscilla, with "crowds and crowds of real happy people having church out in the woods where there were beautiful buildings among the trees . . . every time he had a chance to go to the woods he would look for the place."[13] Brother Wimbish's nephew Jerry Luck recalls his uncle's repeated testimony of the details of his vision:

> A very large place on a hill and the people of God gathering from far and near to worship God in Spirit and truth. He also saw a part was lower farm land with a large house to shelter the saints in time of famine and there was a cemetery to bury the old and poor saints as they pass on from labor to reward.[14]

The Wimbishes moved to Sharon, Pennsylvania, and joined a Baptist church there, all the while still searching for the site he had seen in the vision.

There are several different accounts of what happened next, but they can be easily reconciled. According to Brother Luck, Sister Wimbish sought permission from her Baptist pastor to organize a prayer band that met on Friday evenings and Sunday mornings after service to promote love and forgiveness, prompted by her concern for the hatred that existed among the members. He does not indicate whether this hatred was based upon race. These gatherings led Sister Wimbish and her followers to the experience of Spirit baptism and sanctification: "By much prayer and fasting and studying the scriptures Sister Wimbish was baptized with the Holy Spirit and taught it to us, as a second definite experience which sanctifies all who receive Him and gives us power to live a clean holy life, without which no man can see the Lord."[15] For some unexplained reason, the men in the prayer band were expelled from the Baptist church but not the women. In her own testimony of the group's origins, Sister Wimbish clearly states that she left of her own accord, following what she perceived to be the voice of God:

> We saw the need of a closer walk with God so we started a little prayer band and called it, "The Brothers and Sisters of Love." Those who wanted more of God in their lives became members of the prayer band. We earnestly prayed and studied our Bibles. God revealed the light of His word to us, saved and sanctified us.
>
> God spoke to me later and said, "Come out." I did not know what the voice meant but I obeyed and came out with about seven who followed me. We continued our services in homes and on the streets. I know this was of God because souls were saved and with His help we built a meeting house where the present Church of God now stands. The Sisters and Brothers of Love had no church name and no other church connections. We had never heard of this movement we now know as the Church of God.[16]

It may be that the men were disfellowshiped because they were being led by a woman. A third account of the expulsion is given by Ida Coasy Ard, who was an evangelist in partnership with Sister Wimbish: "Brother Wimbish and Brother Luck were put out of the church; Sisters Wimbish and Luck, and Brother Roddy went out with them."[17] So it seems that some were expelled, while others left voluntarily. In any case, Sister Wimbish was unhindered by whatever actions the Baptist pastor may have taken against her or her male followers and took her prayer band ministry door to door and to the streets of Sharon, Pennsylvania.

A classic testimony of sanctification was given by William P. Blackburn, an early convert won by the prayer band, who eventually became the first pastor of the Church of God in Sharon:

> I finally decided to stop claiming to be a Christian until I could live free from sin. I moved to Farrell, Pa., in the fall of 1909 carrying a burden of sin which

was heavier than before, yet down in my heart I wanted to live for God. One Saturday night in the spring of 1910 I was deeply troubled about my soul and asked God for help. It was about noon the next day I saw a group of men I knew standing on the corner of French and Hamilton Ave. As I joined them one asked me what I thought about living a holy life. I said I didn't know but I was interested in finding out for myself. While I was speaking one of the men said, "There goes one of those holy women in your house now, Blackburn." I left immediately to hear what they had to say. When I arrived they had introduced themselves to Wife and she introduced them to me. They were Sister Wimbish and Sister Coasy (Ard) of the Brothers and Sisters of Love. . . . It was from the lips of these two women that I received a Bible knowledge of the power of God to save and sanctify in this present world.[18]

Brother Blackburn also testified that he made a covenant with the Lord to throw away his prescription medicine and on July 28, 1910, was healed of a stomach and liver condition. He and his wife raised five children without medicine, physicians, or hospitalization.[19] While testifying on the streets of Sharon, Brother Blackburn was called to the ministry.

These experiences demonstrate that this group of black saints sensed that they were being led by God and invested with divine authority. Their collective testimony of divine guidance and spiritual discernment formed the criteria for accepting or rejecting advice and influence from others and especially from whites.

The Sisters and Brothers of Love received guidance from both white and black ministers during this period. The early interracial influences upon this group are apparent in this account of the decision to affiliate with the Church of God, as remembered by the first manager of the camp meeting, Brother Joseph Crosswhite:

One day one of our white brethren stopped in one of their services and after listening he told them, "You people are teaching the doctrine of the Church of God, whose headquarters is at Anderson, Indiana." This was their first knowledge of this Reformation. Afterward they heard of Rev. R. J. Smith, pastor of the Church of God in Pittsburgh, who later came to them and taught them the way of the Lord more clearly.[20]

Grant Anderson, a white Church of God minister, invited the black saints to join the nearby Emlenton Church of God camp meeting instead of starting their own, indicating a willingness, at least on his part, to work interracially. Brother Anderson also introduced the group to R. J. Smith, a black minister he met at Emlenton who had just moved from Freeport, New York, to pastor a Church of God congregation in Pittsburgh. The black saints accepted Brother Anderson's encouragement to become a part of the Church of God but rejected his advice to join the Emlenton camp meeting.[21] In her recollection of these events, Sister Ard did not identify the Church of God ministers by race but, rather, as men and women sent by God:

> In 1912 the Lord sent some Church of God ministers to us, Brother J. L. Williams, J. G. Anderson, R. J. Smith and Brother and Sister Pye. It was through these ministers we received light on the church and changed our name from the Brothers and Sisters of Love to the Church of God.[22]

The newly formed Church of God congregation appointed Brother Blackburn as pastor, Sisters Wimbish and Ard as evangelists, and Brothers Wimbish, Moore, and Luck as deacons. The egalitarian custom of referring to the saints as Brother and Sister, without regard to whether they are ordained clergy or pastors, was not peculiar to the Brothers and Sisters of Love prayer band but was and is widely practiced within the Church of God, especially in the black congregations. In a pattern often repeated in the formation of black churches in the Sanctified movement, the fledgling group was led and housed by women, and, once the congregation was established, a man was called as pastor.[23] Sister Wimbish first received the baptism of the Holy Spirit and led the group to "come out" of the Baptist church, and they first worshiped in the home of Sister Hattie Roddy. On becoming officially incorporated as a congregation, however, the first pastor called to serve was Brother Blackburn. Eventually, they built a meeting house on Cedar Avenue in Sharon. Eighty years later, the congregation would be led by a clergy couple, Pastor Michael Smith with his wife, the Reverend Felecia Smith, serving in an associate capacity, both ordained ministers with graduate seminary degrees.

After the Church of God congregation was established in Sharon, Brother Wimbish persisted in seeking the fulfillment of his vision for a campground for the saints. Brother Smith, who had become a spiritual advisor to the congregation, "caught the vision" and began looking for a campsite in the Pittsburgh area. The site was found by one of the saints, J. A. Christman. While hunting in the woods of West Middlesex, he came upon a hill that reminded him of the place Brother Wimbish had often mentioned in his testimonies. When Brother Christman took him to the site, he confirmed that, indeed, it was the place he had seen in his dreams.[24]

The saints purchased 127 acres of land and, in August 1917, held their first camp meeting. Men and women worked together in constructing the camp meeting facilities and in leading worship. One woman in particular showed a remarkable range of abilities in this regard: Sister Nelson of

Pittsburgh cut trees with the men; helped to install windows, door frames, and roofing; and prayed the first prayer in the first camp meeting.[25]

In the beginning, they called their organization the Western Pennsylvania and Eastern Ohio Camp Ground Association and later changed the name to the Gospel Industrial Association of the Church of God Evening Light.[26] "Evening Light" is a term associated with the Church of God reformation movement in the early years. The "Evening Light Saints" assumed a prophetic identity with reference to Zechariah 14:7b (KJV): "It shall come to pass, that at evening time it shall be light."[27] In selecting a name for their camp meeting association, the saints were very concerned not to create the impression that their work was a split from any state, national, or international ecclesiastical body.[28] The name West Middlesex Campmeeting signifies location rather than exclusive racial identity. Indeed, from its inception, the camp meeting has welcomed the participation of whites and others.

The resolve of the saints to maintain an open, interracial fellowship at the West Middlesex camp meeting was tested and strengthened by the painful rejection some experienced during those years at the international camp meeting convened at Anderson, Indiana. Blacks had been attending this predominantly white gathering for a number of years, and black ministers such as Abraham Stroud of Alabama and evangelist J. D. Smoot were featured as camp meeting speakers. The Church of God General Assembly, a governing body of ministers that is convened each year at the international convention in Anderson, passed a resolution in 1915 to form separate congregations "so that the whites may win more whites and the blacks may win more blacks."[29] Mother Laura Moore, an early convert to the Sisters and Brothers of Love, along with her husband, Samuel, has left a tearful account of her experience at the Anderson camp meeting:

> Some of us have been to the International camp meeting quite a number of times; we had placed it next to heaven. But the last year I was there quite of few of our people came from the south and we were so happy, but a few of the Brethren called for us to meet them. We went to the appointed place and this is what they told us: "There are too many of your people coming here. You'll hinder the whites from coming and being saved. Why don't you get a place of your own?". . . Our hearts were made sad and many tears were shed, for we had no place to go. But later we heard of the meeting here and again our hearts were made to rejoice that we had a place where we could assemble and praise God.[30]

Mother Moore is remembered as one of the saints who worked for many years to "keep a door open" at West Middlesex to allow people to assemble and worship God together regardless of race or color.[31]

The West Middlesex camp meeting was not convened as a response to the racism of the Anderson leaders; Brother Wimbish had received his vision years before he knew the Church of God even existed, and the camp meeting was fully operational before the black saints began to experience alienation from the predominantly white Anderson gatherings. This history presents an interesting case study of the black Christian experience of exile. It would seem providential that a home had already been prepared for the "exiled" blacks before the whites decided to discourage interracial fellowship at Anderson. Not only did they have somewhere to go but also on their own turf, so to speak, they rejected the racist social ethics of the white "saints" and assumed the morally consistent position of refusing to exclude others on the basis of race. Instead of evoking further exclusionary practices, the experience of racism gave them greater light on the importance of maintaining an open door. So, if there is a distinguishing ethical dimension of this oral history, it is not that the saints established a "black" camp meeting but, rather, that they sought to implement an ethics of inclusion consistent with their understanding of Christian love. In time, the West Middlesex campground came to be known as Zion's Hill, a spiritual home and a place of refuge recalling the biblical imagery of exile and return.

William J. Seymour and the Azusa Street Revival

William J. Seymour, the apostle of the Azusa Street Revival that engendered the international Pentecostal movement, is an important figure in the racial dynamics and denominational issues inherent in the emergence of the Sanctified church movement. His life story reflects practically all major facets of the denominational racism experienced by black Christians in the United States. Seymour was born in 1870 in Centreville, Louisiana, to parents who had been slaves, as historian William Montgomery notes, "amid poverty and the intensifying racism of the post-Reconstruction South."[32] In this syncretistic environment, he was subjected to many formative influences, including the distinctively African worship practices of emancipated Christian slaves still living in the plantation setting and the Louisiana Creole religion, which emphasized supernaturalism and Haitian *vaudou*.[33] He was raised as a Baptist, but as a young man he migrated to Indianapolis, where he joined a local black congregation of the Methodist Episcopal Church. He rejected the black Baptist and Methodist churches after his first contact with the Holiness movement, sometime between 1900 and 1902, while he was living in Cincinnati. Pentecostal historian H.

Vinson Synan described the circumstances surrounding Seymour's attraction to the Church of God (Anderson, Indiana) during those years:

> Accepting the Holiness emphasis on entire sanctification, Seymour joined the Church of God Reformation movement, also known as the "Evening Light Saints.". . . While Seymour was in Indianapolis, he contracted smallpox, which left him without the use of his left eye. While reflecting on his illness, he accepted a call to preach and in a short time was licensed and ordained a minister of the "Evening Light Saints" movement.[34]

Montgomery further highlighted the impact of the Church of God on the vision Seymour would promote later as leader of the Pentecostal movement, noting that the interracial group Seymour had joined avowed the entire sanctification of believers, as the apostles of Christ had been infused with the Holy Spirit on the day of Pentecost. They believed that sanctification was more than personal or spiritual holiness, having social or community manifestations as well: "Seymour thought those beliefs, the presence of women in the pulpit, and interracial worship signified the pentecostal movement's commitment to equality and justice in a multiracial society."[35] He was drawn to the Evening Light Saints in the midst of a personal quest for sanctification and physical healing. Seymour's preaching style became more closely aligned with the interracial Holiness tradition than with the black folk tradition: "His speaking ministry was not in the tradition of black pulpit oratory but was more that of a teacher."[36]

Seymour continued his relationship with the Holiness movement after moving to Houston in 1903. There he attended a church pastored by a black woman, Lucy Farrow. In 1905, she went to Kansas to work as a "governess" in the home of Charles Fox Parham. In that same year Parham, a white Holiness preacher who ran a Bible school in Topeka for missionaries, came to Houston to offer a ten-week training session for missionary evangelists. When Seymour enrolled in Parham's classes in Houston, he was subjected to the indignity of having to sit in a hall where he could hear the classes through the doorway, in keeping with the Southern system of racial segregation. Seymour accepted Parham's advocacy of tongues-speaking, but rejected his racist prejudices and polemics. In the meantime, Seymour had been installed as pastor of an Evening Light congregation in Houston, but because the saints did not approve of Seymour's new doctrinal interest in tongues-speaking he left the pastorate.[37]

Seymour was invited by Neely Terry, a Holiness woman from Los Angeles, to pastor a new Church of the Nazarene congregation in California, which had been founded by Julia W. Hutchins. However, Hutchins rejected his preaching and his message that speaking in tongues was the necessary evidence of the Pentecostal experience and locked Seymour out. He found refuge in the home of Richard and Ruth Asberry on Bonnie Brae Street, where he conducted several weeks of prayer meetings. The Asberrys and Hutchins had been expelled from the Second Baptist Church in Los Angeles for embracing Holiness teachings.

When Seymour finally manifested the tongues-speaking experience he had promoted in his preaching, a revival broke out, and crowds began to gather at the Bonnie Brae Street residence and in the streets. He leased a vacant building at 312 Azusa Street in Los Angeles from the Stevens African Methodist Episcopal Church (where several persons worshiping with him had formerly been members), a two-story wooden structure located in a poor black neighborhood in Los Angeles, near some stables and a lumberyard. Inside, a space was cleared large enough to seat approximately twenty persons on wooden planks resting on nail kegs. Within a few days, more than a thousand persons were trying to enter the small forty by sixty-foot mission, and the Azusa Street Revival was underway. The core group consisted primarily of black female domestic workers, but over a period of three years, from 1906 to 1908, the revival drew persons of every race, nationality, and culture.[38]

Seymour was severely ridiculed by the local press within the first few days of the revival because of the tongues-speaking and the interracial worship. On the morning of Wednesday, April 18, 1906, the *Los Angeles Daily Times* published a story that portrayed the Azusa Street gathering as a "new sect of fanatics." Seymour is described as "an old colored exhorter," despite the fact that at the time he was in his mid-thirties. The account is full of racial, sexual, and religious stereotypes:

> The bounds of reason are passed by those who are "filled with the spirit," whatever that may be. . . ."You-oo-oo gou-loo-loo come under the bloo-oo-oo boo-loo" shouts an old colored "mammy," in a frenzy of religious zeal. Swinging her arms wildly about her she continues with the strangest harangue ever uttered. . . . One of the wildest of the meetings was held last night, and the highest pitch of excitement was reached by the gathering, which continued in "worship" until nearly midnight. The old exhorter urged the "sisters" to let the "tongues come forth" and the women gave themselves over to a riot of religious fervor.[39]

When Parham visited the Azusa Street Mission at Seymour's invitation in October 1906, he responded to what he witnessed there in similar terms and called the revival a "darky camp meeting."[40]

Parham's own practice was to make segregated altar calls when he conducted revivals, with whites on one side and blacks on the other. He promoted a white supremacist theology, which held that whites were descended from the Adamic race, and that the flood was God's punishment for the sin of race mixing.[41] He refused to conduct the great union revival Seymour desired and was barred from the Azusa Street Mission after repeated attempts to correct and control Seymour and his followers. Shortly thereafter, Parham lost credibility as a Pentecostal leader when he was arrested in San Antonio, Texas, on a charge of sodomy.[42] He spent his later years as an avid supporter of the Ku Klux Klan. He praised its members for their "fine work in upholding the American way of life."[43]

Racial conflicts with other white Pentecostals further jeopardized Seymour's "dream of an interracial Pentecostal movement that would serve as a positive witness to a racially segregated America."[44] Two white women, Clara Lum and Florence Crawford, helped him to publish the periodical *Apostolic Faith* which had an international circulation of 50,000 subscribers. They were opposed to Seymour's 1908 marriage to Jennie Evans Moore and effectively destroyed Seymour's publication outreach ministry by taking both the periodical and the mailing list to Portland, Oregon, where Crawford founded the Apostolic Faith evangelistic organization. An additional conflict erupted between Seymour and William H. Durham over the "finished work of Calvary" teaching that sinners received both pardon and cleansing upon conversion, against Seymour's understanding of sanctification as a necessary second work of grace. Seymour locked Durham out of the Azusa Street Mission after his attempts to usurp authority on the basis of this new teaching, and Durham's followers established the white Pentecostal denomination Assemblies of God in 1914. Synan concludes that "the struggles with Parham, Crawford, and Durham effectively ended Seymour's major role of leadership in the Pentecostal movement."

By 1914, Azusa Street had become a "local black church with an occasional white visitor." As an apparent consequence of his negative experiences with white Pentecostals, Seymour revised the doctrines, discipline, and constitution of the Pacific Apostolic Faith movement to recognize himself as "bishop" and guarantee that each successor would always be "a man of color."[45] However, after Seymour's death on September 28, 1922, a woman of color assumed the leadership of the mission—his widow, Jennie Seymour. The building was demolished in 1931, and the land was lost in foreclosure in 1938, two years after her death.[46] The land was used as a city parking lot for more than fifty years, and in 1982 a Japanese-American Cultural Center was built across the entire 300 block of Azusa Street.[47] Evidently, none of the Pentecostal churches took action to preserve the Azusa Street site for historic purposes, which can perhaps be interpreted as a further sign of racial hostility and disregard for Seymour's contribution and legacy to the international flowering of Pentecostalism. Tinney reports that a white Pentecostal denomination had refused to buy it for a historic memorial: "We are not interested," they said, "in relics."[48]

Although Seymour's vision of interracial unity in the Spirit may have been repudiated and undermined by some white racist Pentecostals, his work was significantly affirmed and carried forward by Bishop Mason, founder of the Church of God in Christ. After spending a month at the Azusa Street Revival, he returned to Memphis, where he convened all night meetings for a period of five weeks. The white press in Memphis reported the revival with the same racially biased skepticism as did its Los Angeles counterpart. On May 22, 1907, the *Commercial Appeal* published a story under the headline "Fanatical Worship of Negroes Going on at Sanctified Church":

> The pastor pretended to speak the language of the Spirit and the wise ones of the congregation got on to his curves and began using a strange, idiotic jargon, which was alike meaningless to them and the preacher. . . .
> The minister would exclaim "Hicks, hicks!" and the congregation would answer back, "Sycamore, sycamore, sycamore!" and such insignificant words, which lifted the congregation to the highest point of ecstasy, showing what has been contended for years, that the Negro's religion is sound instead of sense.[49]

Despite the negative press coverage, thousands of blacks responded to Mason's spiritual leadership. By 1917, 800 delegates representing twenty states attended the church's tenth annual convocation in Memphis, which was barely noticed by the local press.[50] By the 1990s, the ranks of the Church of God in Christ had grown to 3.7 million, surpassing several of the mainline Protestant churches in total membership, including the Presbyterian Church, U.S.A. (2.9 million), the Episcopal Church (2.4 million), the United Church of Christ (1.7 million), and the American Baptist Churches in the U.S.A. (1.6 million).[51]

Whereas white racism may be the abiding constant in the history of African American Christianity, at least in the Sanctified church movement William J. Seymour has left a legacy of racial openness. A point-by-point comparison of the September 1906 edition of Seymour's periodical *Apostolic Faith* with *What the Bible Teaches,* a compilation of early writings by F. G. Smith, editor of the Church of God publication *Gospel Trumpet,* reveals striking similarities.[52] The only clear point of disagreement is the tongues doctrine. Moreover, the two documents are virtually in complete accord with regard to the two distinguishing doctrines of the Church of

God, holiness and unity. It appears that Seymour maintained his Church of God orientation toward doctrine and Scripture; therefore, he can be seen as a tongues-speaking Church of God preacher. In other words, Seymour embodies the hyphen in Holiness-Pentecostal. His commitment to Christian unity had a dual grounding, in the antisectarianism of the Church of God reformation movement and in the Pentecostal blessing of tongues-speaking as a ritual reenactment of the outpouring of God's Spirit on all flesh. Seymour believed that the revival itself was a sign of divine approbation of racial unity, having more to do with "the crossing of racial lines and the achievement of unity in the body of Christ than the formal criterion of glossolalia as proof of spirit baptism."[53]

The Apostolic Faith Mission was vitally linked to the Holiness movement via the doctrine of sanctification and the practice of accepting as equals women and men of all races, but, in time, the tongues-speaking issue and the persistence of white Christian racism presented insurmountable obstacles to Christian unity. Although it may be true that the Azusa Street Revival was shepherded by an ordained Church of God preacher, the doctrine and practice of speaking in tongues were never widely accepted in the Church of God, and Seymour is hardly ever mentioned and certainly not "owned" by Church of God scholars. Nevertheless, the dialectic Seymour employed in his dealings with whites remains instructive for subsequent generations of black Christians who share his commitment to interracial unity – openness to whites who are accepting of the offer of fellowship and repudiation of whites whose racism and other moral failings produce alienation and strife within the church.

Participation of Women and the Poor in the Holiness-Pentecostal Movement

The original Azusa Street congregation had at its core a group of black female domestic workers, a group typically overrepresented in the ranks of the Sanctified churches at a time when the vast majority of black women were employed as domestics. In Seymour's own words, "The work began among the colored people. God baptized several sanctified wash women with the Holy Ghost, who have been much used of Him."[54] As has been demonstrated in the case of the West Middlesex camp meeting, women ministers and leaders played a formative role in the Holiness movement during the early twentieth century. The Holiness movement had attracted significant numbers of blacks, beginning in the late nineteenth century, and the first black congregation in the Church of God was founded and pastored by a woman named Jane Williams in Charleston, South Carolina, in 1886.[55] However, just as the Holiness and Pentecostal churches tended to compromise their early stance for interracial unity by splitting and segregating by race, there were corresponding adjustments made in the leadership roles of women.

In 1978, Pearl Williams-Jones surveyed five major Pentecostal bodies and categorized them with respect to their treatment of women's ministry and leadership.[56] The first category, consisting of churches who insist on the subordination of women in ministry roles, actually comprises the overwhelming majority of black Pentecostals: the Church of God in Christ, the Church of Our Lord Jesus Christ of the Apostolic Faith, and the Bible Way Church of Our Lord Jesus Christ World Wide. The second category, churches that grant women positions of authority equal to men, includes the Pentecostal Assemblies of the World and the Mount Sinai Holy Church of America (which was founded by a woman, Bishop Ida Robinson). The one church body surveyed by Williams-Jones that was founded by a woman, the Mount Sinai Holy Church of America, lists Mary E. Jackson as presiding bishop in the *Directory of African American Religious Bodies*.

Church historian Susie Stanley uses the term "stained-glass ceiling" to describe increasing barriers to women's leadership and advancement in denominations with a long history of ordaining them. For example, women in the Church of God are not substantially represented among the top tier of agency executives who serve the national body.[57] Stanley cites statistics showing the decline in the proportion of women clergy in the Church of the Nazarene from 20 percent in 1908 to 1 percent more recently, and a less precipitous decline in the Church of God (Anderson, Indiana) from 32 percent in 1925 to 15 percent. Other Holiness churches that ordain women show similar patterns of decline, with the sole exception of the Salvation Army, where women today still constitute a majority of commissioned officers, or ministers, as they did a century ago in 1896. Stanley claims that, compared with mainline denominations that began ordaining women only in recent years, the Holiness movement has a "usable past," having recognized women in ministry dating back to the mid-nineteenth century.[58] Women in five Holiness denominations—Church of God (Anderson, Indiana), Church of the Nazarene, Free Methodist Church, Salvation Army, and the Wesleyan Church—currently constitute 25 percent of the clergy in their denominations, whereas women comprise 7 percent of the clergy in thirty-nine other denominations that now ordain women.[59]

With respect to the issue of class, Leonard Lovett has described the Azusa Street Revival as a grassroots movement, a contribution "from the ghetto to the world" and "symbolic of a microcosm of pentecostal and ethnic ecumenicity."[60] Although the Azusa Street Mission attracted a multiracial multitude of black, white, Hispanic, Native American, and Asian seekers of both sexes and all social classes, the fact remains that its fundamental identity as a group of poor black women and men significantly facilitated the revival's broad appeal. However, as long as congregations of believers within the Sanctified church tradition continue current trends toward increased educational levels, higher income, and greater affluence, they will remain challenged to sustain a quality of ministry that welcomes the poor into creative, productive fellowship with believers of all economic classes. Tinney has observed that "more money turns over in the Pentecostal churches on any given Sunday than circulates through all the other Black businesses during an entire week," but "these churches have little to show for their millions, except devotion to the American ethos of materialism."[61] The next chapter describes the approach taken by one contemporary Holiness congregation to dismantle the barriers of race, sex, and class in its worship and to promote reconciliation in solidarity with the urban poor.

2

Refuge and Reconciliation in a Holiness Congregation

The story of a modern urban Holiness congregation is presented here to serve two purposes. First, the overview of a local church whose congregational history spans the greater part of the twentieth century should help to corroborate at least a few of the general insights and issues presented in this study as characteristic of the exilic motif in African American religious life. Second, this account illustrates some of the practical concerns and challenges engaged by pastors of Holiness-Pentecostal people whose worship and work is informed by the call to be saints—"in the world, but not of it."

Pastoral Leadership at Third Street Church of God, 1910–1995

The Third Street Church of God had its earliest origins in the Christian witness of a family who migrated to Washington, D.C., from Charlotte, North Carolina, during the first decade of the twentieth century: Sister Minnie Lee Duffy; her brother, Elder James E. Lee; her sister, Sister Viola Lee Cyrus; her mother, Sister Cherry Lites Lee Johnson; and her aunt and uncle Brother and Sister Doc Lites. This first Church of God mission in the nation's capital was established in 1910 in a small room in the home of Sister Cherry Lites Lee Johnson on Six and One-Half Street, Southwest. They held church in their home and invited ministers passing through Washington to speak to their small but growing congregation. On one such occasion, Elder Charles T. Benjamin, a traveling evangelist based in New York, was invited to return and subsequently became the shepherd of that small flock.[1]

Elder Benjamin was born on October 17, 1878, on the island of St. Kitts in the British West Indies. He attended school there, furthered his education in private study, and was saved and called to the ministry as a youth. He desired to go to Africa to do evangelistic work and decided to come to the United States to acquire financial support for that ministry.

Elder Benjamin came to the United States in 1900 and was able to find employment. He saved his money and soon was on the first leg of his journey to Africa. His first stop was London, England, where in his hotel room God revealed to him the inadvisability of going to Africa alone without backing from any person or church group. He returned to the United States and opened a mission in New York, where he served as pastor and leader of a group of devoted followers.

One day, while walking along Eighth Avenue in New York, his attention was attracted to a mission, over the door of which was the caption Church of God. He was greatly impressed with the mission's doctrines of holiness and unity and, shortly thereafter, became affiliated with the Church of God. While living in New York, Elder Benjamin wrote *The Fall of David's Tabernacle, Rebuilt by Christ,* an overview of Old Testament themes in light of the accomplishments of Jesus Christ. Just prior to receiving the call to pastor the mission in Washington, Elder Benjamin married Minnie Lee Sweitenberg of Newberry County, South Carolina. She was a close friend of Sister Duffy, who spearheaded the pastoral call.

As the congregation grew, they established a mission home in 1913 at 914 S Street, Northwest, in the Westminster section of the Shaw community. Many well-to-do black families resided in the area, and there were numerous thriving black-owned business establishments nearby.

The three-story brick rowhouse served several purposes: as a parsonage for the pastor and his wife, as housing for saints and newcomers to the area, and as a facility for worship and Sunday school. Here the Benjamins provided preaching, teaching, housing, employment assistance, and general support for students, single adults, and young married couples in a Christian atmosphere. Under the leadership of the Benjamins, the 914 S Street Mission acquired a reputation, not only in Washington but also throughout the United States, especially the South, as a place where persons coming to Washington for jobs or education could receive assistance and support. They required cleanliness, discipline, respect, and shared responsibility of the residents of the mission home. However, it is clear that all these strict rules and regulations of the household were enforced and followed in a context of Christian love. Residents did not live there free of charge, and proceeds from the rental of rooms were used to defray the expenses of the church. Their style of pastoral ministry can perhaps best be described as communal parenting.

There were close connections between parental and pastoral ministry at the 914 S Street "church home." In addition to weekly worship services held there on Sundays, Wednesdays, and Fridays, family prayer and Bible study were held at least twice daily, beginning at 6 A.M. A Sunday school department was established, with classes for adults, youth, and children. A Women's Missionary Society was founded with Sister Benjamin serving as president. Elder Benjamin gave leadership to the church's evangelistic ministry by conducting street meetings and tent meetings throughout the city, sometimes featuring guest evangelists.

From 1928 to 1947 the congregation numbered seventy-five to a hundred persons, composed of a core group of fifteen or so families. They remained a closeknit, family-oriented fellowship after relocating to a new building at 1204 New Jersey Avenue in 1928. With the exception of the pastor, no one was paid for services rendered: musicians, custodians, and secretaries all volunteered their time and energies to serve the needs of the congregation. Ministries of visitation and outreach continued to grow in various forms, including Elder Benjamin's summer tent meetings in the far Northeast section of the city. He formed relationships with well-known ministers of other groups in the Sanctified church tradition, including Bishop Samuel S. Kelsey of the Church of God in Christ, Elder Lightfoot Solomon Michaux, founder of the Gospel Spreading Church of God, and Bishop Smallwood E. Williams of the Bible Way Church of Our Lord Jesus Christ World Wide, with headquarters located one block south on New Jersey Avenue.

Worship was an experience of sound doctrinal preaching; inspiring congregational, choral, and special music; and exuberant shouting of the saints. In the opinion of many, Elder Benjamin was more of a lecturer than a preacher. His sermons exemplified a profound knowledge of Bible history and culture and were presented in a professional manner. Although reserved himself, Elder Benjamin was quite tolerant of the shouting of his own members, who leaped and ran and cried out as the Spirit moved them during the worship services. However, his attitude toward shouting was summarized in the simple admonition, "Don't shout any higher than you live."

The tradition of discipline and order that characterized life at 914 S Street was continued at the New Jersey Avenue location. Families were expected to sit together in worship, and children were kept in line. The Holiness doctrine was clearly preached, taught, and formulated in terms of specific standards and expectations for Christian living. One member recalls being required to wait five years before being baptized, an indication of how persons were required to demonstrate their Christian commitment prior to partaking of the ordinances of the church.[2] Strict dress codes were enforced, especially for women, who were not permitted to wear beads, earrings, makeup, sleeveless dresses, or shoes exposing the toes.

Sister Benjamin lectured the young women on how to keep house and take care of their husbands and children. Elder Benjamin taught the men to work to support their families and to give priority to their leadership responsibilities at home. Although he was generous in offering financial assistance to families in need, he expected men to be financially responsible at a time when many married women did not work outside the home. Family life was centered on the church and church-sponsored activities; participation in scouting, fraternities, sororities, bowling, skating, dancing, theater going, and ball games was strictly prohibited. Sick members were expected to follow the biblical mandate to "call for the elders of the church" rather than to seek medical attention from doctors and hospitals. Many testimonies of divine healing emerged from that practice.

During the Depression years, the influx of migrants from the South greatly increased, as persons sought jobs, housing, and economic opportunities in the nation's capital. The patterns of migration were such that most of the church members had roots in North Carolina, South Carolina, and Virginia. In the 1940s, both during and after World War II, the concentration of government jobs attracted persons from the South, North, and Midwest. The Church of God advocated conscientious objector status for young men drafted during the war, based on convictions against taking up arms, and the pursuit of such status brought a few young men to Washington and to the church.

One of the most painful events in the pastoral ministry of Elder Benjamin grew out of his encounter with racist attitudes among some of the whites at the Anderson camp meeting prior to 1947. In one of the meetings there, a suggestion was made that all the Church of God congregations should contribute funds toward establishing a national church in Washington, D.C., the sentiment was that "it is a shame that we don't have a church in Washington." Elder Benjamin, who was present at the time, took offense at the fact that the congregation he was pastoring in Washington was completely disregarded in the plan, presumably because the whites did not want to be officially represented in the nation's capital by a predominantly black congregation with a black pastor. He never returned to Anderson. Despite the racial hostility and lack of support from the church headquarters in Anderson, the congregation continued to welcome whites into the fellowship, and Elder Benjamin ordained white ministers.

Although Elder Benjamin understandably lost his enthusiasm for participation in the international camp meetings and other activities at Anderson, Indiana, following the events of the 1940s, he maintained a high profile within the National Association of the Church of God and the annual West Middlesex camp meeting. For many years, he served as secretary-treasurer of the Foreign Missionary Board at West Middlesex, on the Business Committee, and as a member of the Ministerial Assembly.

During the 1950s, a noteworthy effort was undertaken as an expression of compassionate ministry to the aged. In August 1955 the Peeler Benevolent Fund of the Church of God, later known as the Peeler Foundation for the Aged, was founded and directed by Sister Nola Peeler (1889–1987). Her motive was to help elderly persons who suffered because of lack of family or financial resources to provide for their care. The congregation hosted many fund-raising programs in support of the Peeler Foundation and home.

A major milestone in the history of the congregation was the dedication of the main auditorium at 1204 Third Street, Northwest, on November 30, 1947. This sanctuary was subsequently named the Benjamin Memorial Chapel after the pastor's death. With faithful stewardship by the members and under careful fiscal management by the pastor and financial officers, the congregation prospered during the 1950s and was able to burn the mortgage on the Third Street property in May 1959. The church address was changed from New Jersey Avenue to Third Street as a result of a zoning reconfiguration implemented by the city government.

In 1967, as Elder Benjamin entered his eighty-ninth year, age began to take its toll, and his health declined. The late 1950s and early 1960s represented a time of tremendous social upheaval, brought about by changes in race relations, modes of social protest, and the involvement of churches in the civil rights movement. However, Elder Benjamin's concept of ministry did not necessarily reflect a concern for these changes in the society, and the congregation was not actively involved in social justice issues, aside from what individual members and auxiliaries undertook on their own, with or without his endorsement. He retired in 1967 after 57 years of leadership, with the distinction of having pastored one congregation longer than any other minister in the Church of God. Elder Benjamin died one year after his retirement, on August 9, 1968.

After a two-year interim period, Samuel George Hines was installed as senior pastor of the Third Street Church of God on Sunday, September 7, 1969. Like his predecessor, Elder Benjamin, he was a native of the West Indies, born on April 19, 1929, in the town of Savanna-la-mar on the island of Jamaica. His father, the Reverend Dory Alexander Hines (1883–1952), was a Church of God pastor. His mother, Vesper Dell, was a graduate of Tuskegee Institute in Alabama. Despite his family's lack of wealth and social status, Samuel Hines successfully completed preparatory and secondary education in Jamaica. Converted at the age of sixteen and called to the ministry as a young man, he pursued ministerial studies in Jamaica, England, and the United States. While living in London, he established and led an organization known as the International Christian Fellowship. He was ordained to the ministry in 1952, and in 1953 he married Dalineta Louisa Ellis, also of Jamaica.

Prior to coming to Third Street Church, Hines had pastored several congregations in Jamaica, England, and the United States. He took on the particular challenges of bringing his own distinctive pastoral style and convictions to a congregation whose previous leadership and history he held in highest esteem: "I was privileged to come here to succeed a man who had built such a strong church. He built properly a soundly based, stable Church of God congregation. I didn't have to do a lot of laying foundation. I had to take a maintenance-oriented church to turn out a mission-oriented church."[3] Early on, he invested his time and energies in community involvement by working with numerous government agencies, community c:ganizations, and ecumenical groups. Hines also served as the local representative of Operation PUSH (People United to Save Humanity).

Reconciliation was the central theme undergirding the pastoral leadership of Hines. He challenged the congregation to become "Ambassadors for Christ in the Nation's Capital," proclaiming the message of reconciliation as directed by the apostle Paul in 2 Corinthians 5:17–21. In 1972 a ministry statement was developed and adopted to this end:

> We are ambassadors for Christ in the Nation's Capital, committed to be a totally open, evangelistic, metropolitan caring fellowship of believers. To this end we are being discipled in a community of Christian faith, centered in the love of Jesus Christ and administered by the Holy Spirit. We are covenanted to honor God, obey His Word, celebrate His grace, and demonstrate a lifestyle of servanthood. Accordingly, we seek to proclaim and offer to the world a full cycle ministry of reconciliation and wholeness.

As a constant reminder of the congregation's identity, this statement appeared each week in the Sunday worship folders and in many other official documents, pamphlets, and publications of the church. Moreover, from time to time the pastor would call on the gathered congregation to read or recite it in unison. It was often used as a centerpiece for discussion during the annual Journey Inward, a four-day series of meetings held each fall to focus the concerns of the congregation in particular ways. The Journey Inward also served as a forum for reviewing and revising this ministry statement as needed.

Hines expressed his commitment to the ministry of reconciliation in other ways. He gave critical behind-the-scenes leadership to the demolition of apartheid in South Africa by aggressive and focused counseling of church leaders, political activists, and governmental leaders. The church sponsored other members on short-term visits to South Africa to do work projects such as construction and children's ministry. Closer to home, he insisted that the weight of biblical evidence places great responsibilities on the church to minister to the poor. The congregation's energies and resources became focused on bringing the powerful and the powerless together in ministry among the urban poor, guided by Hines's conviction that "Dogma divides and mission unites."

Three major programs were undertaken to give expression to Hines's urban theology and strategy for ministry to the inner city: (1) the urban prayer breakfast, which provides a worship experience and morning meal for people who live in the local shelters and on the streets each Monday through Friday beginning at 7:30 A.M., weekly food package distribution, and special community dinners at Thanksgiving and Christmas; (2) a clothing distribution program; and (3) a tutorial and counseling program that meets daily throughout the year (afternoons and evenings during the school year and all day during the summer vacation) to provide educational support and counseling for thirty community children. The urban prayer breakfast remains the centerpiece of the church's ministry of reconciliation.

The U.N.I.Q.U.E. (United Neighbors Involved in Quality Urban Experience) Development Corporation was spawned by the Third Street Church as an independent corporation to build a model community based on the reconciliation concept. Another organization known as One Ministries was founded as a parachurch group with offices located in the same block as the church. The founder and director of One Ministries was Brother John H. Staggers Jr. (1922–1990), a professor in the Department of Sociology at Howard University who had served as a special assistant to D.C. Mayor Walter Washington from 1968 to 1971.

Throughout his twenty-five years of pastoral ministry at Third Street, Hines was recognized as a gifted preacher and exegete, continuing the tradition of preaching excellence established by Elder Benjamin. Hines traveled extensively as an evangelist; he often preached at conventions, camp meetings, ministers' caucuses, and revivals, both in the United States and abroad. He was frequently featured as a preacher at the annual international conventions of the Church of God in Anderson, Indiana, and at the various gatherings of the National Association of the Church of God. His voice was heard weekly on local Christian radio on such programs as "Word of Reconciliation," "Voice of Reconciliation," and "Word for Washington." His eloquent preaching was matched by his mastery of the written word; his book, *Experience the Power,* was published by Warner Press in 1993. In addition, he produced numerous articles, study papers, and other manuscripts for the benefit of the local church and the wider Christian community. He conducted classes and workshops in homiletics throughout the nation and taught a variety of continuing education classes for pastors.

Hines fulfilled several major leadership assignments with the central agencies of the Church of God in Anderson and with the National Association of the Church of God at West Middlesex. He served from 1983 through 1989 as chairman of the General Assembly of the Church of God, the church's most representative decision-making body of ministers and lay leaders. He was elected or appointed to numerous boards and commissions, including the board of directors and the publication board of Warner Press, the publishing house of the Church of God, as well as the Missionary Board. He died suddenly while in recovery from surgery on January 6, 1995.

A comparison of the pastoral leadership of Hines and Benjamin would reveal many roots and fruits in common but very different roles and challenges. Elder Benjamin's major role was as founder and builder; Hines's emphasis was on expansion and outreach. As a couple, the Benjamins were known for the hospitality and help they gave to migrants from the South and elsewhere; Dalineta Hines, functioning especially in her role as administrative assistant to her husband, joined him in leading the church's ministry to the homeless poor of Washington and in giving support to immigrants from the West Indies through personal contact and special events such as the annual West Indian Reunion.

Over a period of eighty-five years, the Third Street Church embodied a dialectics of refuge and reconciliation under the leadership of these two pastors. Elder Benjamin understood the church to be a refuge from sin and worldly living, a home away from home for persons seeking employment, education, and Christian fellowship in the urban environment. Hines emphasized reconciliation across class and racial lines so that the hospitality of a hot meal and a warm welcome could be extended to refugees of the urban crisis—the homeless, prostitutes, alcoholics, drug addicts, and the unemployed. Both pastors sought to cultivate a sound, spiritual environment for winning souls and nurturing believers in a cosmopolitan city that historically has held forth many attractions and opportunities for persons of other states and nations.

Sunday Worship

The following is a description of a typical Sunday morning of worship at Third Street Church of God, as observed during the summer of 1994, a few months prior to Hines's death. Beginning at 9:30 A.M., small groups of children and adults gather for Sunday school classes in the church's main sanctuary, the fellowship hall, and in an apartment building next door that has been converted into classrooms. These Sunday school classes meet for an hour or so, offerings are quietly collected in each class, and then there is a brief assembly for announcements, prayer, and dismissal in the sanctuary led by the Sunday school superintendent. The level of activity picks up as the choir members put on their robes and gather for prayer and last-minute rehearsing, the sound technicians arrange and test the microphones, and the uniformed ushers greet the worshipers, hand them worship folders, and guide them into their seats. At the same time, the pastor and ministers meet in a small office off the pulpit area, exchange pleasantries, review special prayer requests, put on liturgical robes, pray, and make assignments of who will lead each aspect of the worship as printed on the order of service.

The service begins with an organ prelude at 11 A.M. Within a few minutes, the choir assembles at the rear of the sanctuary with the ministers for the procession. There is a choral introit, "We Are Standing on Holy Ground." Then the pastor reads a call to worship, Psalm 92:1–2:

> It is a good thing to give thanks unto the Lord, and to sing praises unto thy name,
>   O most High:
> To shew forth thy loving kindness in the morning, and thy faithfulness every
>   night.

The processional hymn is an up-tempo selection from the Church of God hymnal; the words of the chorus reflect the heritage of the Church of God, known in the early days as the Evening Light Saints:

> To thy house, O Lord, with rejoicing we come,
> For we know that we are thine.
> We will worship Thee in the Bible way,
> As the evening light doth shine.

While singing, the choir members march in single file, women first and men following. Next, the ministers fall in line, with the pastor coming in last. Once those in the procession have taken their places in the pulpit area, all four verses of the hymn are sung, with choir, ministers, and congregation all singing together, accompanied by organ and piano. Usually, there are no drums or tambourines. One of the ministers offers a prayer of invocation, in which she invites God's guidance and anointing of the ministries of music, prayer, preaching, and praise.

The worship leader comes forward with further exhortations to worship, leading the congregation in singing several choruses and hymns in succession, "What a Mighty God We Serve," "He Is the King," and "Majesty." The congregation is enthusiastic in singing praise music, the simple, melodic choruses whose lyrics typically are derived from Scripture: "We are a chosen generation, a royal priesthood, a holy nation, a peculiar people." Between songs, the worship leader interjects exhortations to praise and invites members of the congregation to greet one another and to repeat some phrase to their neighbors.

Everyone who is able is asked to stand throughout the praise singing—some are clapping, raising one or both hands in worship, waving hands, or swaying to the rhythm of the music. The entire congregation and choir join their voices together in enthusiastic praise, with sporadic shouts of "Amen!" "Hallelujah!" "Glory!" and "Praise the Lord!" Latecomers are still entering the sanctuary, which is filling up with men, women, and children, all ages and both sexes fairly evenly represented. Most of the worshipers are dressed in their "Sunday best"—the men wearing suits and ties, although it is a hot June day, and the women wearing dresses and suits, some with hats. The vast majority are African Americans, but the assembly includes a significant number of persons of West Indian descent, a few Africans, whites, and persons of mixed heritage. The sanctuary seats approximately 250 persons. The official average Sunday morning attendance is 220, and the total church membership is 275.[4]

The worship leader concludes this part of the service with a brief prayer and invites the congregation to take their seats. A minister then gives a welcome to visitors and asks them to fill out a card that will be collected and read toward the end of the service, when the visitors are individually acknowledged.

The pastor now rises to direct the congregation to share special requests for prayer. Usually the minister of visitation gives a brief report on the condition of members of the congregation who are sick at home, in the hospital, or preparing to have surgery. The pastor asks everyone to review the list of sick and shut-in members and friends of the congregation that is printed in the worship folder. An opportunity is given for everyone to raise one hand in the air to signify a personal need for prayer, and a few persons speak aloud urgent requests for family, friends, or neighbors. The worship leader directs a hymn sung in preparation for prayer: "O Victory in Jesus, My Savior, Forever!"

The pastoral prayer begins with praise, adoration, and thanksgiving to God. Then petitions are presented based on the requests shared on behalf of the sick and bereaved. Needs are addressed beyond the scope of the immediate church family. Prayer concerns listed in the worship folder include the Church of God in Barbados and "the neighborhood around the church; for peaceful redemptive relationships among all persons." The pastor names specific areas of the world plagued by war, famine, and other manifestations of human suffering. He prays for the success of the service and concludes by inviting all worshipers to recite together the Lord's Prayer. The choral response to the pastoral prayer is "Praise the Name of Jesus."

Next, the offerings are taken. The pastor makes a few remarks reminding the congregation of their biblical obligation to pay tithes, one-tenth of their total sources of income, each week. The benevolent offering is taken each week to support a fund for financial crises incurred by members and also to support the church's weekday outreach to the poor. Another minister reads an exhortation on giving that is printed in the worship folder, and the ushers come forward to receive the offering plates and pass them row by row from the rear of the sanctuary to the front. While the offering is being taken, the children's handbell choir plays an offertory selectio. Their performance captures the attention of the congregation, who respond with vigorous applause. Once the offerings are collected, the ushers present the full offering plates at the front of the sanctuary, a lively doxology is sung, and one of the ministers offers a prayer of thanksgiving.

A minister reads announcements concerning the schedule of church events for the coming week. A Scripture lesson is read from 2 Corinthians 5:14–21. There is a contemporary gospel selection by the pianist, a talented composer, music director, and soloist who provides his own piano accompaniment to a solo, "We Need a Word from the Lord."

The pastor ordinarily begins his sermon by inviting everyone to stand and greet one another. After offering a brief prayer, he draws attention to the Scripture lesson that has been read aloud or a related text, usually designated as part of an ongoing series of sermons on a theme followed through a particular book of the Bible. The sermon, which usually lasts about an hour, begins with a thorough, detailed exegesis of the text, develops its main body in terms of a three-point structure, and concludes with some sort of exhortation to repentance and/or rededication based upon the divine mandates drawn from the text.

Invariably, the sermon ends with an altar call, and persons who are so moved come forward to kneel at the altar. There may be a great gathering of almost the entire congregation standing around the altar when there is no longer room left to kneel. These altar calls are invitations to prayer, and, while the appeal is being made, the choir and congregation may sing an invitational hymn. Persons who come forward are able to receive personal prayer and brief counseling from a small cadre of altar workers. The pastor leads the congregation in an intercessory prayer offered on behalf of those who have responded at the altar to his appeal.

After the worshipers have returned to their seats, a closing hymn is sung, the congregation is commissioned to serve with words reflective of the sermon and text for the day, and a final benediction is pronounced. The choir and ministers recess to the rear of the sanctuary, the pastor positions himself at the door to greet the departing worshipers, and the congregation disperses. From the call to worship to the final benediction, the total elapsed time is approximately three hours, more on the first Sunday of the month, when the Lord's Supper is observed. People spend considerable time greeting one another, and there are at least two brief meetings of church auxiliaries in the sanctuary for a short time following the conclusion of the service.

The Urban Prayer Breakfast

At 6:30 A.M. on a Monday morning during the summer, a senior husband and wife team who are members of the Third Street Church arrive to begin preparing the food for the weekday urban prayer breakfast. Both are in their eighties, but very active and energetic; they volunteer their services on a weekly basis. They are joined by four men who assist in cooking the grits, eggs, sausage, and toast. Additional volunteers descend the seven or eight steps down from the street level into the church fellowship hall on the basement level, and they help by pouring juice, cutting pastries, setting up coffee, and arranging chairs.

By 7:15 A.M. the ministry team has arrived, and they begin setting up the sound system and synthesizer on the stage. Their responsibility is to lead the worship, which consists of songs and prayers, and to provide a brief sermon concluded by an invitation to discipleship. At that time, the kitchen crew drops everything for a moment to huddle in the kitchen for prayer; three black women, three white men, and two black men hold hands in a circle and pray for God's blessing on the ministry that is about to begin. A group of twenty-five white high school students from North Carolina arrives before 7:30 A.M. They are affiliated with Young Life, an evangelical youth ministry organization, and will be spending an entire week in the city of Washington. The students approach the persons in charge to receive their individual assignments to help with serving the food. Everything runs very smoothly and efficiently.

The worshipers take their seats in the 200 or so metal folding chairs arranged in tight rows facing the stage, which is actually a platform built over the baptismal pool. It is not easy to distinguish the volunteers from the "poor" in that all are casually dressed in T-shirts, shorts, and jeans, and the persons who come in from the streets are not disheveled or inebriated. Of the 160 who are served breakfast on this day, at least 90 percent are young black males. There are fewer than a dozen women, almost all black, and some of the men appear to be middle-aged or elderly. The forty or so volunteers include three members of the church; the ministry team composed of persons from three local charismatic congregations including blacks, whites, Cambodians, and Ethiopians; the all-white Young Life group of teens accompanied by several leaders who are college-age and above; and five or so black men closely identified with those being served, except that they obviously have roles in assisting in the food service and securing the entrances and exits. The daily prayer breakfast program depends heavily upon volunteers, and the only advertisement is by word of mouth among the persons on the streets and in the shelters. There is always a diverse mix of people present, men and women of all races, including all ages, with the exception that children rarely are present. On the average, 150 to 200 persons gather each weekday for worship, work, and fellowship around a substantial "soul food" breakfast, a term aptly applied to both the menu and the ministry.

The worship begins at about 7:30 A.M. The praise team offers contemporary inspirational music—brisk praise choruses and Scripture songs from the white charismatic movement, such as "Come into This House," "Mourning into Dancing," "He Is Exalted," and "Celebrate Jesus." The synthesizer provides complete accompaniment, keyboard chords, solo instrumental voices, and the rhythmic beat of the percussion instruments. The singing is led by a black woman, two black men, and a white man. The seats facing the stage are now mostly filled, and latecomers are directed to fill in the empty seats scattered from front to back. The audience is attentive and responsive but seems not to know most of the songs. However, they join in the singing, occasionally lifting their hands, standing on their feet, or saying "Amen." A few of the men in the congregation sleep, read the newspaper, or otherwise tune out the worship without distracting the others.

After thirty to forty minutes of singing, one of the men on the ministry team steps forward to begin preaching. The sermon is not delivered in any rigid, formal manner; the preacher is as casually dressed as the others, and he engages the men with questions, challenges, and observations but not in a condescending manner. At several points in the message, he exhorts them to seek the kingdom of God, to choose Jesus Christ as Savior, and to acknowledge that poverty is no excuse for failing to follow the Lord. The message ends with a prayer and an appeal for the men to come forward for prayer. None do, perhaps because it is 8:30 and time for breakfast to be served. An hour later, however, after almost everyone has left, one man who is obviously troubled and weeping approaches the ministry team for help, and they respond by quietly laying hands on him and praying.

The senior man in charge stands by the doorway to the kitchen with breakfast tickets in hand. They are distributed and collected to make sure that no one doubles back to receive seconds. He calls for "ladies first," and only three women come forward. Then the men fall in line, beginning with those seated in the first row and ending with those seated in the back. They are very orderly, cooperative, and patient. Within forty-five minutes, everyone has been served. The portions are substantial, and there is enough food for all. The ministry team continues to sing while breakfast is served, but people are free to engage in their own conversations as they eat. Most people leave as soon as they are finished eating, and there is a volunteer who attends the exit door. No one is allowed into the building after 8:30, and there is a volunteer posted at the entrance to secure it after breakfast has begun. He turns away several people because they are too late. After everyone has been served, the man in charge invites the volunteers to come through the line. The last to be served are the members of the ministry team, who seem to want to keep singing even after most people have left. On some days, one of the ministers leads a Bible study group that is open to everyone but draws perhaps a dozen regular participants.

The spirit of the gathering is extremely congenial, and one senses a distinct orientation toward mutuality, humility, and service. Among the volunteers, most are eager to greet and meet each other informally. The worship leader routinely introduces any groups who are visiting as volunteers, but they mingle on their own among the men and among other volunteers. After the breakfast has ended and the cleanup has begun, one of the Young Life leaders talks enthusiastically about how the urban prayer breakfast has been a life-changing experience for her. When asked to explain, she says that it gave her a new vision of the family of Jesus Christ, something she had never seen before in her own church or community. She has come to Washington each summer for the past four years, first as a teen and now as a college senior. She is excited about the local evangelical leaders who will be meeting with her group during the week, and she has been favorably impressed by the pastor of the Third Street Church, who has been a guest preacher at her church in North Carolina. Most of the other volunteers seem genuinely happy to be present and involved in the urban prayer breakfast and feel challenged to share the gospel of Jesus Christ among the urban poor.

The ecumenical and interracial nature of the ministry is striking. The volunteers come from many churches and denominations; some are local and others come from various regions of the United States. It seems easy for them to put aside doctrinal, liturgical, racial, and cultural differences in order to come together with Washington's poor to lift up the name of Jesus.

3

"In the Beauty of Holiness": Ethics and Aesthetics in the Worship of the Saints

Basic Elements of Sanctified Worship

There are numerous articles, books, and dissertations that directly describe worship in the Sanctified church and others that rely on James Baldwin's masterful description of the conversion of his protagonist, John Grimes, in *Go Tell It on the Mountain* (1952) as their major source of information about the rituals of the movement. James Shopshire (1975), Arthur Paris (1982), and Joseph Murphy (1994), have written descriptive narratives of black Pentecostal worship based on participant observation. The content of these three narratives are analyzed in this chapter in conjunction with the black Holiness worship described in chapter 2, in the process of assessing some general areas of similarity and difference between the Sanctified church and mainline black Protestant churches in terms of worship.[1]

Shopshire's 1975 doctoral dissertation in the field of sociology of religion includes a lengthy description of his Sunday morning visit to a black Pentecostal church in Chicago. The outline of events is

1. devotional, a practice derived from the Southern rural tradition, including Scripture reading, a cappella singing of "old songs," chanted responsive prayer led by a deacon, and organ prelude;
2. procession of clergy and choir, with the choir marching and swaying from side to side;
3. introit and chant of the Lord's Prayer;
4. responsive Scripture reading, followed by another hymn and prayer of blessings;
5. songs of praise and inspiration, with worshipers making verbal acclamations, standing, clapping, walking the aisles, and dancing;
6. reading of announcements;
7. reading names of sick and shut-in members;
8. welcome and recognition of visitors;

9. sermon beginning with Scripture read by a ministerial elder, the preacher using lively illustrations, encouraging responses such as standing, waving hands, shouting words of approval, smiling, crying, holding trance-like gazes, speaking in tongues, and making antiphonal climax with organ accompaniment;
10. response and altar call as traditionally done in most Black Protestant churches, with invitation to repentance first, an invitation for new members to join the church next, followed by a second altar call with the preacher praying for all who come forward;
11. offerings, with tithers (those who give 10 percent of income to the church) given special appeals and visibility;
12. the holy dance, which has been assigned "special significance" in the national body with which the congregation is affiliated;
13. benediction and words of praise and satisfaction with the "good time" that had been experienced.

Shopshire commented that the first three elements are much the same as in any Protestant worship gathering, "with the probable exception that the singing was better than average." Moreover, what he referred to as the "formalized order of service" comprises everything up to and including the responsive Scripture reading. Shopshire noted in his narrative that the Sunday service begins at 11:45 A.M. and ends at 3:00 P.M. The particular service he has observed may be somewhat atypical in length, however, in that it included the service of Communion (which in many black churches occurs only on the first Sunday of the month) and a live radio broadcast (which was actually an abbreviated recapitulation of the entire service of worship).[2]

Arthur Paris's *Black Pentecostalism* (1982), a book that is a revision of his doctoral dissertation in sociology, provides a summary description of the typical Sunday morning worship at three Boston congregations of the Mount Calvary Holy Church of America, Inc. He organizes his outline of the service in terms of three general categories: devotional service, service of the Word, and closing. Paris lists twelve component elements that are very similar to what Shopshire recorded:

1. opening song(s), given as a signal for people to "stop chatting and to settle down";
2. Scripture reading by the devotional leader, who commonly chooses a selection from the Psalms;
3. requests for prayer, which are announced by members of the congregation;
4. prayer, offered by the devotional leader and accompanied by vigorous vocal and hand-clapping responses, with a hymn softly hummed as background prior to and during the prayer, for example, "I Need Thee Every Hour;"
5. song service, with hymns sung from hymnals or from the oral tradition, during which people may shout and get happy;

6. testimony service, with members giving conventionalized testimonies following a basic three-step pattern (hymn sung with congregation joining in, words of acknowledgment and thanks, and a request for the continued prayers of the saints), the devotional leader closing with a standardized formula (e.g., "If there are no further testimonies, we will now bring this part of our service to a close and will now turn the service into the hands of our pastor") and the congregation rising as the minister comes forward;

7. offering(s), with members marching to the front of the church to place their money in offering plates on a table, a procedure that may be drawn out until the desired amount is obtained;

8. choir selection(s) sung in preparation for the sermon;

9. sermon, described by Paris in threefold form of (a) prayer, Scripture reading, and announcement of subject; (b) explication of subject with parables and analogies; and (c) climax and call to sinners to accept Jesus Christ as Savior;

10. altar call, an appeal to sinners to repent, with the preacher conversing with and praying over persons newly saved and having them testify, which may include prayers for healing and laying on of hands;

11. announcements and recognition of visitors, done by another minister;

12. benediction and dismissal by the preacher, who raises both arms and pronounces a trinitarian blessing and amen.[3]

Joseph M. Murphy, a historian of religion, has included a detailed description of a Church of God in Christ worship service in Washington, D.C., in *Working the Spirit* (1994), his survey of five diasporic religious traditions. His narrative can be summarized as

1. call to worship, with a church mother leading the Negro spiritual "Everytime I Feel the Spirit" from the pulpit after the choir and deacons have taken their places without a formal procession;

2. recitation of the official denominational creed;

3. chanting of the Lord's Prayer and "Yes, Lord," which Murphy cites as the Church of God in Christ national anthem;

4. general altar prayer of intercession, led by an elderly church mother, with the entire congregation kneeling in their pews;

5. Scripture readings from the Old and New Testaments, read "with authority" as call and response between lector and congregation, who respond by clapping, raising hands, and shouting "Amen," "Yes," and "Praise the Lord;"

6. special selection by the choir and soloist, with the congregation responding by shouting and speaking in tongues;

7. anointing service, during which persons come to the altar for healing, a small quantity of olive oil is smeared on their heads, and the pastor lays hands on them, assisted by nurses and deacons;

8. sermon, in which key Scripture passages are read and repeated, accompanied by "organ bursts, drums, handclaps, and shouts;"
9. ministry of tithing, with the pastor soliciting, in terms of specific amounts (a hundred dollars, twenty, ten, five, and so on), donors, who come forward and put their donations in a wooden box, and at the end the pastor asking everyone, including those who do not have any offering to give, to form a single file in the central aisle and move forward toward the altar where the box has been placed (the money counted and receipts signed by the ushers in front of the congregation);
10. final benediction.[4]

The Third Street Church of God worship service described in the previous chapter can be recapitulated here in outline form for comparison's sake:

1. organ prelude, choral introit, and call to worship led by pastor;
2. processional by choir and clergy;
3. prayer of invocation led by assistant minister;
4. praise choruses and hymns sung by the congregation and choir;
5. pastoral prayer of intercession, which follows the solicitation of special prayer requests and congregational singing of a meditational hymn;
6. recitation of the Lord's Prayer;
7. offerings, with baskets passed in the pews by the ushers and an instrumental offertory selection, ending with the singing of the doxology;
8. announcements;
9. Scripture reading;
10. special musical selection in preparation for the sermon;
11. sermon with scattered verbal responses from the congregation;
12. altar call, an invitation to salvation and discipleship, during which an invitational hymn is sung and ministers pray for individuals kneeling at the altar;
13. closing hymn;
14. commission to serve and final benediction offered by the pastor.

An outline of the typical Afro-American church service was used as a structural form for the jazz composition "In This House, on This Morning," recorded and released in 1994 on compact disc by trumpeter Wynton Marsalis. In this composition, he uses the idiom of blues and jazz to give interpretative voice to the black worship experience. Marsalis, the son of a New Orleans jazz musician, was not "raised" in the church. He solicited the worship outline from a pastor-scholar with firsthand knowledge of the black church tradition in its full ecumenical and spiritual diversity, Jeremiah Wright, who is senior pastor of the Trinity United Church of Christ in Chicago. The typical structure can be summarized in twelve parts: (1)

devotional, (2) call to prayer, (3) processional, (4) representative offerings, (5) hymn, (6) Scripture, (7) local announcements, (8) altar call, (9) sermon, (10) invitation, (11) recessional, and (12) benediction. The composition actually takes the form of an expanded version of this outline, including the Lord's Prayer, an up-tempo postlude, and a fellowship dinner.[5] Melva Costen's *African American Christian Worship* (1993) discusses some of these same elements, but her worship outline seems intended as a corrective model based on mainline Protestant practices.[6]

It is not difficult to reconcile and condense the four narrative descriptions of worship into a composite portrait of worship in the Sanctified church tradition. They all have at least eight basic elements in common, with some variation in order: (1) call to worship, (2) songs and hymns, (3) prayer, (4) offerings, (5) Scripture reading, (6) preaching, (7) altar call, and (8) benediction. The recitation of the Lord's Prayer and the reading of announcements are found in three of the four. The worship outline for the typical black Protestant churches includes a few more elements than most of the others, adding the recessional, processional, and invitation, but its structure is not substantively different from the composite outline of the Sanctified church worship in terms of the key worship events. Important factors that account for some of the variation among these descriptions are the degree of familiarity and understanding the participant observer brings to the study of worship and the application of different sets of categories and labels to the same or similar phenomena. For example, what Shopshire and Paris cite as "altar call" is probably the same thing as "invitation" in the black Protestant outline. What Murphy designates as an "anointing service" may be essentially similar to the "altar call" at the other churches. Where possible, the description of each of the composite elements is made in broad enough terms to encompass variations in nomenclature and interpretation.

The call to worship includes acts that initiate the worship experience. It may be a simple and informal verbal signal to "stop chatting and settle down" or a formal combination of choral introit and litany recited by minister and congregation. The call to worship may be a brief reading from the Bible, the church's hymnal, or some printed worship aid that encourages people to become focused on worship. In some cases, it is preceded by a devotional service, including songs, prayers, and testimonies. Also, it may be immediately followed with a processional by the clergy and choir. In the church Paris described, the devotional service comprises half of all that happens in the entire worship experience, if not also half of the total worship time. In the church Murphy depicted and in the Third Street Church of God, there are no formal devotions as such. In all cases, some verbal signal is given to invite the congregation to worship.

The singing of some combination of songs, hymns, choruses, and Negro spirituals is a vital part of all these worship services. It is difficult to denote the role music plays in worship with any degree of precision because music tends to undergird everything else that is done. Unlike some of the other elements of worship, music is interspersed throughout the service and not at just one or two points in the order of worship. In the composite outline, however, the singing of songs and hymns represents a major component of congregational involvement in the worship experience. The sacred repertoire is inclusive of hymns of the mainline evangelical Protestant church, gospel songs, praise choruses, and Negro spirituals. Shopshire, a United Methodist clergyman, seems given to understatement when he judged that the worship in the Pentecostal church he observed is much the same as "any of the Protestant denominational worship gatherings, with the probable exception that the singing was better than average."[7] The Sanctified church is known for its enthusiastic singing and response. Moreover, it is a living repository of a full range of black sacred musical forms. The sung repertoire of the Sanctified church tradition includes classical anthems, arias, oratorios, hymns, gospel songs, spirituals, shouts, chants, and lined-out common-meter sacred folk songs.

Bernice Johnson Reagon, director of the program in African American Culture at the Smithsonian Institution and veteran musician of the civil rights movement and the singing group Sweet Honey in the Rock, seemed genuinely perplexed by the shouting she witnessed in response to classical sacred music at Tindley Temple in Philadelphia, a historic black Methodist church with a strong Holiness tradition:

> As the service unfolded that June morning, I was able to witness a cultural tradition that embraced both the order and selections of well-loved "high" church literature and the practice, richness, intensity, and spontaneity found in the most traditionally based Black forms of worship. There were hymns, anthems, prayers, and creeds. There were "amens" and hand-claps and shouts of "Thank you, Jesus" and a spirit that ran throughout the service. . . . They talked about Reverend Charles Albert Tindley as a man who believed in Holiness, believed in tarrying services where people prayed and sang all night long, believed in involving members from all walks of life in the service in whatever way they needed to be spiritually fed.
>
> These beliefs produced a worship form in which members would shout "holy" after [Rossini's] "Inflammatus," and that simply did not compute for me.[8]

At the Third Street Church of God, there are three keyboard instruments in the sanctuary, a Steinway grand piano, a Conn console organ, and a Hammond organ, which enable the performance of a broad range of liturgical music. Notwithstanding the diversity of music that can be used in Sanctified worship, there is a distinctive sound associated with the Sanctified churches produced by musical instruments such as the Hammond organ, piano, keyboard synthesizers, drums, tambourines, bass guitar, and saxophone. The history of the Hammond organ is significant because it was patented in 1934 and manufactured in Chicago at about the same time that gospel music began to be performed in the Chicago black churches. Kenneth Morris, the organist at First Church of Deliverance who introduced the Hammond organ to Chicago and to the world in 1939, has described its special qualities:

> On a Wurlitzer, you had the ordinary tones, solo, vox humana [stops that imitate pipe organ sounds, although the tone actually is produced by reeds]. There was also the slowness of the action [speech, typical of reed sounds]. It was preset, and you could do nothing with it but use the preset keys [stops]. With the Hammond, of course, you had preset keys, but you also had the whole range of your imagination you could play with— you could play a dog whistle if you knew how to do it. The action was precipitous, instantaneous, just like a staccato is on the piano, and you could make it as legato as you wanted. It's endless what can be done with a Hammond organ.[9]

The Hammond organ became the instrument of choice for the improvisational style of worship music in the Sanctified churches. Murphy paid special attention to the importance of the organ in the worship service, noting at many points in his narrative the manner in which the organ shapes the mood and expresses the energy of the songs, speech, and dance. The organ takes the lead in providing the rhythmic and tonal texture of the worship experience, and it is the principal instrument used to accompany the chanted sermon. Both Murphy and Shopshire described the call and response between preacher and organist, which is actually a three-way conversation involving preacher, congregation, and musician. In the hands of a skilled and accomplished musician, the organ sings, speaks, and dances.

Prayer is an individual or collective appeal to God, which includes praise, thanksgiving, confessions, and various petitions. As is the case with music, it is difficult to fix one point in the outline of worship at which prayer occurs because it typically is done repeatedly throughout the service. Prayers are sometimes chanted in the Sanctified church, in a manner not unlike the chanted sermon. They are seldom read or recited from a printed source, with the exception of the Lord's Prayer, which the worshipers may recite or chant from memory. In all but one of the worship experiences described here, the Lord's Prayer is spoken or chanted at some point in the service, representing a vital ecumenical connection with the prayer rituals of the universal church.

Offerings are taken by having the worshipers march to the front of the sanctuary to deposit their monetary gifts for the church in baskets, on plates, or on a table. Also, the ushers may pass the offering receptacles up and down the rows of seated congregants in a precise, orderly fashion. The Sanctified churches emphasize tithing, and sometimes special prominence is given to the tithers by having them come forward individually to place their tithes in a special receptacle. The offerings can consume a considerable amount of time if the minister makes an appeal for a specific cause or if people are asked to bring their offerings according to the specific dollar amount, as is the case in the churches described by Murphy, Paris, and Shopshire. Usually some form of prayer and/or doxology is offered in connection with the offering, either before or after the monies are actually received.

Scripture reading is another indispensable element in Sanctified church worship. One or more texts may be read near the beginning of the service or shortly before the sermon is preached. The Scriptures can be individually read from the pulpit or read responsively by minister and congregation. The Bible is accorded the highest respect and regard in these churches, and in some cases there are special ritual procedures for transporting and handling the particular Bible from which the sermon is preached.

Preaching is a climactic event in Sanctified church worship because it is believed that the preacher actually speaks for God. Often the sermons in the Sanctified churches are performed in the sense that the basic message and content are amplified through chants, moans, dancing, and other ecstatic behaviors. Each of the worship narratives presented here describes the interaction between preacher and congregation in multiple dimensions. Preaching is more than the simple verbal communication of the gospel of Jesus Christ based upon some scriptural text; it involves emotion, physical movement, and various modulations of the preacher's voice and is designed to bring the worshiping community into some form of climactic expression—shouting, tears, praise, repentance, tongues-speaking. In some of the churches, specific provision is made for the preacher (typically male) to have an attendant (typically female), whose responsibility is to assist him with his liturgical cape, to administer juice or water as needed, to wipe the sweat from his brow, and so on, adding to the dramatic impact of the preaching performance. Sociologist Harold Dean Trulear has described the ritual aspects of preaching in this context: "The use of robes, capes, etc., to enhance the preacher's appearance and the attendant nurse with her ever-present orange juice and fresh handkerchiefs are all part of the props or staging of the ritual drama where 'God speaks to His children.' "[10]

Regardless of the size of the sanctuary, these churches all have electronic sound systems, some of them very sophisticated and advanced, and the preachers use handheld and/or lapel microphones to enhance the modulation of the preaching voice. The sermon is always intended to elicit congregational response.

Altar call is a formal ritual of response to the preached word, which usually functions as an invitation to discipleship. Many Sanctified churches adhere to the practice of issuing dual altar calls—the first an appeal for sinners to repent and receive salvation and the second an invitation for believers to receive sanctification or the baptism of the Holy Spirit. Altar calls may also include the ritual laying of hands upon the sick or distressed and anointing with oil, with the purpose of achieving healing or deliverance. Prayer is always a key element of this ritual, which may occur at some other point in the service, even prior to the sermon. In some churches, the major objective of the altar call is to invite the worshipers to have hands laid on them so they can be "slain in the Spirit." The dissociative experience of temporary loss of consciousness represents a form of ritual empowerment. In summary, the altar call may serve a variety of purposes in worship. It is used to invite sinners to repentance, new converts to church membership, hurting persons to wholeness, and saved persons to sanctification and other forms of spiritual empowerment and blessing. For some worshipers, the altar ritual is as pertinent and significant to them personally as the sermon itself, if not more so. There are preachers who invest as much time and energy in directing the altar call as in preaching the sermon.

Benediction is a prayer or formula of blessing that signals that the worship experience has ended. It may include a final exhortation or commission of the worshipers to implement some particular truth or principle that has been preached. The minister who offers the benediction may raise one or both hands, and in some cases the worshipers also raise their hands while receiving the benediction.

One additional comment is in order with regard to the comparison of the composite Sanctified worship structure with its typical black Protestant counterpart. In his jazz recording "In This House, on This Morning," Marsalis intentionally employed a broad range of sacred sounds and elements drawn from the global religions of African peoples in his interpretation of the "typical" black church service. However, the particular rhythms and vocalizations in the "Holy Ghost" climax of the selection entitled "Sermon" are clearly recognizable as Pentecostal. Thus, the composition conveys the important idea that the "typical" black Protestant church worship may bear the specific stamp and influence of the Sanctified church tradition.

Saved, Sanctified, and Spirit-Baptized

As used in this discussion of worship, "saint" is a term suggestive of both liturgical and ethical identity. The key testimony or confessional formula that characterizes the saints is "saved, sanctified, and filled with the Holy Ghost." Each denomination among the Holiness, Pentecostal, and Apostolic churches has specific doctrines and disciplines governing the interpretation of the meaning of salvation, sanctification, and spirit baptism, but some generalizations are ventured here in an attempt to describe the liturgical and ethical self-understanding of the Sanctified church as a whole.

To be saved means that one has repented, asked forgiveness of sins, and confessed Jesus Christ as Savior and Lord. This experience imparts a basic "entry level" of liturgical identity that distinguishes the saint from the unbeliever. To be sanctified is to receive some second form of blessing that conveys upon the believer a distinctive ethical identity of being set apart for God, literally to be made holy. Some of the non-Wesleyan groups would not see sanctification as a separate process but as an experience inherent in salvation. To be filled or baptized with the Holy Spirit is a declaration of liturgical identity that signifies that the saint has experienced total initiation into the worshiping community by a personal confession or manifestation of spirit possession. The evidence for this experience is the major area of doctrinal difference that accounts in part for the vast multiplicity of denominations and church bodies within the Sanctified tradition. Generally speaking, the Wesleyan-Holiness churches emphasize the infilling of the Spirit as manifested in a holy life, whereas the Pentecostal and Apostolic churches seek the pouring out of the Spirit in the ecstatic utterances of tongues.

James Tinney, who testified that he "got the Holy Ghost" during his adolescence in 1956, offered a vivid portrayal of the experience of tarrying for spirit baptism in the black Pentecostal context:

> So the seeker prays loud and long as hard and as fast as he can to get this power. He sweats and cries and screams and physically throws himself, demanding that God do what he wants. He commands the power of God as his own. It is a violent scene—one which is carefully hidden from the casual visitor. The seeker will work himself into a state of possession if it takes hours upon hours of struggling. Hair will become matted, clothes will become dirtied, the flesh will become sick and feint until "the power comes." . . . The result will be a total rejection of American mainstream values, coming back full circle to the African heritage of possession. And it will be symbolized by a break with rational thought and language and an utterance in unknown tongues, among other manifestations.[11]

FIGURE 1. A sister does double duty catering the Urban Prayer Breakfast at Third Street Church of God . . .

FIGURE 2. . . . and singing as a member of the worship team.

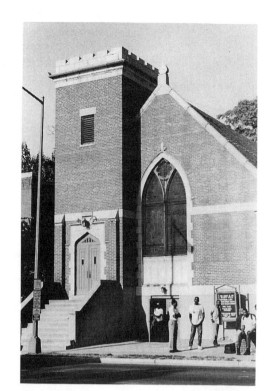

FIGURE 3. Worshippers from the streets of Washington linger outside the church after breakfast ends.

FIGURE 4. Altar call during Sunday morning worship at Third Street Church of God.

FIGURE 5. The pastor anoints the head of a worshipper for healing as the assisting minister lays on hands.

FIGURE 6. Panoramic view of the Church of God Campgrounds at West Middlesex, Pennsylvania.

FIGURE 7. Ushers dressed in white form a line outside the tabernacle in preparation for worship.

FIGURE 8. Song writer and recording artist Bill Gaither addresses the campmeeting crowd from the tabernacle pulpit.

FIGURE 9. Worship on "Zion's Hill."

FIGURE 10. The saints rejoice in a hotel ballroom during the National Inspirational Youth Convention of the Church of God.

FIGURE 11. Liturgical dancer.

FIGURE 12. A
child recites before
the congregation.

FIGURE 13. Church of God Children's Choir sings at Bible Way Church.

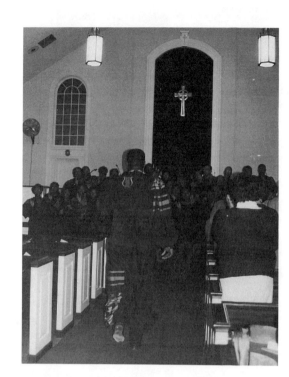

FIGURE 14. Howard
University Gospel Choir.

FIGURE 15. Swarthmore
College Gospel Choir.

Tinney's account is a part of his political science doctoral dissertation; as would be expected, his understanding of spirit baptism is couched in the language of power. Moreover, the social-ethical focus of his interpretation entails a ritual return to Africa and a concomitant rejection of American mainstream values, presumably both religious and secular. It is important to acknowledge that this ritual is conducted in secret, or at least removed from the purview of the casual observer, as Tinney suggested.

All of the Sanctified churches acknowledge water baptism as a necessary act of ritual obedience that identifies the believer with the church of Jesus Christ. Some churches invoke trinitarian formulas for baptism and others unitarian, that is, baptizing in Jesus' name. Unlike the rituals of spirit possession, water baptism is a very public ritual. Both the Sanctified and black Protestant churches in the rural South have historically practiced water baptism outdoors in creeks and rivers, and some of the major urban churches have sponsored massive baptismal services in public places in the cities.

Although baptism by immersion is ordinarily required only once in the life of the believer, the baptism of the Spirit may be understood as a ritual of initiation that can be repeated, replenished, or reenacted as often as the saint becomes possessed by the Holy Spirit in worship. The fact that the possessing Spirit is holy mandates that the saints manifest holy living both inside and outside the sanctuary. Thus, there is a vital connection between the ethical and liturgical identity of the saints, as expressed in the exhortation of the Psalmist: "Rejoice in the Lord, O ye righteous: for praise is comely for the upright" (Psalm 33:1, KJV).

Static and Ecstatic Forms of Spirit Possession

Spirit possession is an important feature of virtually all the diasporic religions of New World Africans. For example, devotees of Cuban *santería*, Haitian *vaudou*; and Brazilian *candomblé* enact elaborate rituals of possession and acknowledge a corresponding pantheon of possessing spirits and deities of African derivation. What separates the Sanctified church tradition from the others is the belief that the possessing spirit bears the exclusive identity of the Holy Spirit.

The perennial objective of Holy Spirit possession is achieved in some combination of ecstatic and static forms. A trance resulting from religious fervor is the salient feature of ecstatic worship forms. *Ecstasy* literally means "out of place." In static worship forms, by contrast, worshipers are at rest or in equilibrium. *Static* literally means "causing to stand." However, as in describing electricity, static worship should not be assumed to be necessarily dead or lifeless. Static electricity is electrical force produced and accumulated as potential energy; current electricity moves and flows in the form of kinetic energy. Static energy is stored; kinetic energy moves. The two are interdependent because kinetic energy is the discharge of static energy through some conductor or channel. Yet, as anyone who has observed a thunderstorm can attest, static energy can be discharged at random, without conductor or channels, with a force that is not only impressive in magnitude but also frightening and potentially lethal in impact. By contrast, the flow of kinetic energy can be entirely predictable and controlled. The two distinct forms of electrical energy suggest an analogy that can bring enhanced insights to the study of Sanctified worship, a dialectic of static and ecstatic worship structures and forms of spirit possession.[12] Along the continuum of Holiness-Pentecostal groups, the Holiness churches tend to favor the static forms and ideals of spirit possession, and the Pentecostals insist upon ecstatic expression in worship.

To define ecstatic worship as worship "out of place" necessitates formulating some understanding of its dialectical opposite, worship "in place." Static worship is the state of equilibrium out of which the ecstasy flows; it is the requisite platform for the trance ritual to occur. In no way should this scheme be understood as indicative of the relative inferiority or superiority of static and ecstatic forms of worship. The fact remains, however, that practically every Christian worship tradition tends to favor one over the other, sometimes to the exclusion of the other. Yet, the two are neither equal nor mutually exclusive in this sense; ecstatic experience absolutely depends on static structures, but static structures may or may not produce ecstatic experience. In fact, static forms and structures can be intentionally used to deny or suppress spirit possession. The ecstatic state may be forthrightly suppressed, scorned, or forbidden. So there can be static worship without spirit possession in any state or form, but it is not possible to have worship of any kind without some static structure to initiate and organize the ritual interaction of the worshipers. In other words, to say that static structures sustain ecstatic forms of worship is merely to agree that one cannot dance without a floor, sing without a scale, or speak without a language.

Static structures are those elements in worship that represent a state of equilibrium or rest. With reference to the composite outline of Sanctified worship, they include hymn singing, Scripture reading, corporate prayers (especially the Lord's Prayer), offerings, sermon, altar call, announcements, and benediction. These structures are designated here as static because they embody the potential energy of the worshiping congregation to explode into ecstatic expression: shouting or holy dance, tonguesspeaking, spontaneous utterances, and lifting holy hands. Most of these structures can serve as a platform for ecstatic movement. For example, people may shout during hymn singing and sermons, speak in tongues during or after prayer, and engage in call and response as the Scriptures are read. The call to worship and the benediction are also static structures that frame the worship ritual by marking the boundaries of sacred time. Generally speaking, the offerings and announcements do not support ecstatic activity or evoke ecstatic response.

Thus, worship has fixed and fluid forms, rehearsed and unrehearsed, scripted and improvised, and prepared and spontaneous. To make matters more complex, it is clear that some forms and events in worship reflect both fixed and fluid elements at the same time. For example, the quintessential ecstatic expression in Sanctified worship is the shout, or holy dance, which usually occurs as a spontaneous eruption into coordinated, choreographed movement. There are characteristic steps, motions, rhythms, and syncopations associated with shouting. It is not a wild and random expression of kinetic energy. Rather, a culturally and aesthetically determined static structure sustains the expression of ecstasy in a definite, recognizable form, the existence of which may not be apparent to the casual or uninformed observer. Similarly, speaking in tongues may appear to be a strictly spontaneous and unrehearsed verbal expression, but, in reality, the practice is evoked by "tarrying" or other repetitive patterns of activity designed to encourage tongues-speaking. Glossolalia is not the only ecstatic speech used in worship. The vocabulary of utterances spoken spontaneously in worship is not random or undefined. There is a definite lexicon for intelligible ecstatic utterances in the sanctuary that may manifest cultural and regional variants but is, nevertheless, known to the group. Most of these terms can be found in the King James Bible with reference to the praise and attributes of God: "Hallelujah," "Amen," "Glory," "Holy," "Praise the Lord," "Yes," and "Thank you Jesus." In the ecstatic state, the worshiper may repeat one or more of these expressions many times, in a loud or subdued voice.

Among black Protestant churches in general and Sanctified churches in particular, there are two basic orientations toward worship that set the tone for worship in particular congregations: quietistic and lively. The quietistic congregations give priority to static structures; the lively congregations value ecstatic expressions in worship. Quietistic worship traditions may exclude or control ecstatic worship forms by, for instance, insisting that everything in worship be scripted, read, and timed; by restricting rhythm and repetition, especially in singing; or by direct intervention or verbal rebuke by authorized figures such as ushers or preachers. Lively worship traditions may devalue static worship forms by making statements such as

"We are not here for form or fashion, we are here to praise the Lord" or by vigorously exhorting persons to speak aloud, stand, raise their hands, or shout and subjecting them to verbal ridicule if they refuse, as in "You think you're too cute and too sophisticated to shout." The quietistic worship leader imposes silence and stillness upon the congregation; the leader of lively worship invokes noise and motion. Interestingly, the task of setting the tone for worship, whether quietistic or lively, is not always totally determined by the minister, singer, or preacher who is standing in the pulpit; leadership may be exercised indirectly, but to great effect, by the one who organizes, reproduces, and distributes the order of worship or by some individual or group in the congregation to whom the worship leader looks for cues and approval. Then again, the congregation as a whole may be predisposed to one or the other style of worship, and collectively, by their silence or their utterances, they may indicate approval or disapproval of what is taking place. For example, if one individual is given to loud utterances in a quietistic congregation, the response may be staring, frowns, or hushing actions. In a lively congregation, however, the individual who prefers to remain still and silent may feel uncomfortable and self-conscious and even attract unwelcome public criticism or ridicule. In his study of black worship, Trulear defended the integrity of quietistic worship in black middle-class churches as a legitimate ritual verification of a particular concept of humanity:

> If being human means to be dignified and intellectual, under control and logical, all patterns of behavior that this society has said Blacks are incapable of, then these congregations will model these ideas of human virtue in the context of worship. This is still a function of Black humanity. Those who would deny this as in some sense legitimate would have to eliminate people such as Du Bois and Daniel Payne from the Black religious world.[13]

In this perspective, it is helpful to bear in mind that competing ideals of black humanity may be at stake in debates between lively and quietistic worshipers concerning appropriate forms and expressions of black worship.

Static and ecstatic worship have their distinctive sets of gatekeepers. Ushers, nurses, and deaconesses—that is, uniformed attendants with some designated title and role—are the gatekeepers of the static aspects of worship. Singers, preachers, and, to some extent, dancers are the gate-keepers of ecstatic worship, who "usher" the congregation into and out of the ecstatic state. Ushers attend to the physical movement of worshipers in and out of the sanctuary and demarcate the temporal and spatial boundaries that encompass the sacred space. In other words, as ushers greet and seat each worshiper, they are defining and managing the ritual space; their tenure of duty spans the entire worship time, from prelude to benediction. The preachers and singers direct the emotional and spiritual dynamics of the worship experience, and ushers participate in this process by attending to the special needs and security of persons experiencing the transition from static to ecstatic worship.

With respect to gender roles in general, the Sanctified churches tend toward a peculiar egalitarianism in assigning gatekeeping roles based on gender. The gatekeepers of static structures can be men, just as the gate-keepers of ecstatic expression—the preachers and worship leaders—can be women. Both men and women serve as lead singers and dancers, according to gifts and ability.

For obvious reasons, the persons chosen as ushers, nurses, and attendants tend not to be easily and readily inclined to ecstatic spirit possession. Similarly, the other gatekeepers, including preachers, singers, and instrumentalists, are normally expected to know and honor the rules governing the static forms and structures of worship and to maintain spiritual equilibrium whenever the congregation is swept into the ecstatic state. The biblical principle invoked as an explanation for the need for gatekeepers to maintain equilibrium is taken from 1 Corinthians 14:32, Paul's letter addressed to an early charismatic Christian congregation: "And the spirits of the prophets are subject to the prophets" (KJV).

Worship and Exile

Given that the ultimate objective of worship in the Sanctified church tradition is some form of spirit possession, the aesthetic and ethical norms that govern movement toward this objective are derived from the Bible and black culture. The distinctive songs, speech, and dances of the Sanctified church symbolically "usher" the saints "out" of this world and into a more authentic one discerned within sacred time and space. What makes this process exilic is the connection made between the saints' rejection of the world and the world's rejection of the saints. The saints reject the world on the basis of biblically derived ascetic commitments, that is, the mandate to holiness; they are themselves "rejected" by the dominant host culture because of their race, and sometimes their sex and class. What is not so clear is whether there is consensus among the Sanctified communions that the mandate to holiness dictates opposition to racism, sexism, and economic oppression. When the saints sing "Holy" unto the Lord, lift up holy hands, or do the holy dance, in effect, they are expressing their allegiance to a world in which God has determined who is accepted and who will receive power. Moreover, their worship shows that they believe God accepts the praise, performances, and aesthetic standards that are characteristic of Africans in diaspora. The Holy Spirit has freed at least some of them from the pressure to conform to the worship styles of the dominant culture. The saints are "in" a world that is sinful, oppressive, and discriminatory; they demonstrate that they are not "of" this world by purging themselves of its secularizing influences through rituals that meet their own criteria for cultural authenticity and biblical interpretation. In worship, the saints replicate the "other" world, the place where the exile can be at home. Ethically, their allegiance to this other world requires them to be loving, honest, and pure, even in relations with their enemies. Just as the sanctuary or temple is the place of ritual possession, their bodies are the temple of the Holy Spirit. Ritual purity in the sanctuary requires purity of body, mind, and spirit outside the sanctuary. By their worship, the saints manifest the holy character of the God they serve. By clean living, the saints demonstrate to the world and in the world that they possess the Spirit that possesses them in worship.

In view of these various descriptions of worship in the Sanctified church tradition and the suggestions made thus far with regard to how these data may be interpreted, it is now possible to identify some of the specific marks and symbols of exilic worship, that is, elements and events in worship that distinguish these churches from the white North American Protestant mainstream. This list of four is hardly exhaustive: (1) the holy dance, (2) the "Yes" chant, (3) the use of white-uniformed liturgical attendants, and (4) the inclusion of community announcements and the welcoming of visitors as liturgy. Many or most of these marks and symbols can be found in traditional black denominational churches and definitely in "neo-Pentecostal" Baptist and Methodist congregations. To characterize these facets of worship as "exilic" is to suggest that they are rooted in African cultural identity and may be reflective of specific worship patterns and cultural practices associated with slave religion in the rural South. Each is defined and illustrated in the remainder of the chapter from the data provided thus far, and their specific ethical and aesthetic meaning within the community of the saints is explained. As needed, references are made to additional visual data and performance notes not already cited in the summary outlines.

The Holy Dance: Shouting

The holy dance is best exemplified as the ritual of the shout, the climactic expression of individual and collective spirit possession that is especially characteristic of the black Pentecostal congregations. In her article "Dancing to Rebalance the Universe: African American Secular Dance and Spirituality," Katrina Hazzard-Gordon has commented that dance serves as a "kinetic vocabulary" through which the needs, perceptions, impressions, and responses of African American people are articulated.[14] Two of the distinct types of steps Hazzard-Gordon designates as "conduit steps" in rituals of spirit possession are specifically related to the liturgical dancing of the Sanctified church tradition: the "resting step," a simple shifting of the weight from side to side, from one foot to the other with a slight twisting away from the center of movement, as observed in the swaying motion of black gospel choirs; and the "shout step," a simple, two-footed rhythmic hopping up and down, often with the feet never leaving the floor.[15] Her description of the juxtaposition of the "chaotic, the uncontrolled, and the unconscious" movements associated with the onset of full possession with the "ordered, contained, conscious, and controlled" conduit step is reflective of the static-ecstatic dialectic in Sanctified worship.[16]

In this perspective, the concept of liturgical dance can be expanded to include choreographed choir processions and a whole host of bodily gestures by choir and congregation, such as swaying, patting of feet, clapping of hands, raising one or both hands, and spontaneous standing up. In the Holiness churches, there are saints who do not do the "shout step" associated with the Pentecostals but rather leap straight up and down when they "get happy." In addition, skilled liturgical dancers in some of the Sanctified churches perform carefully choreographed and rehearsed dances to various types of sacred music. At the Third Street Church of God, the liturgical dance troupe, Praise in Motion, consists of young women who perform arranged dances to spirituals, gospel songs, and contemporary praise music. Although executed with great energy and enthusiasm, these liturgical dances are not ecstatic but would be more properly described as exemplary of the static form.

In his worship narrative, Shopshire gave some indication of the aesthetic and ethical norms the saints associate with the holy dance. His account is illustrative of the tension that sometimes exists between the static and the ecstatic in fulfillment of the expectations of the worship leader:

Not being satisfied with the response, [the bishop] said in a scolding tone, "I can't understand how anyone can remain quiet and seated in such a spirit-filled gathering as this. Get up, and dance!" Speaking especially to the constituent members of the gathering, he took time to remind them that to dance is indicative of a meaningful experience in worship, and they "need not try to be cute" by not talking back and dancing. . . . As he talked he was moving back and forth across the length of the pulpit platform with a very agile gait, ever so often initiating a brief dance step and then stopping. By the time the point had been made about dancing being integral to meaningful worship experience he had reached a vocal peak, and performed a dancing frenzy for about 15 seconds.[17]

Clearly, this bishop has mastered the technique of inciting the holy dance through measured demonstration. He seems to have a definite sense of the aesthetic requirements of the ritual dance. Moreover, he seeks to convince others of the ethical propriety, even necessity, of ecstatic expression in worship.

Church historian Winthrop Hudson's 1968 article "Shouting Methodists" confirms the notion of the shout as exilic liturgy. Hudson argued that "shouting" was a conspicuous feature of early Methodism but was not restricted to the Methodists or particularly indigenous to the United States. He cited the ring shout as a form of corporate praise with a lack of inhibition and restraint. Eventually, the shouters were "exiled" from the Methodists ranks because some Methodists were unhappy with the camp meeting type of worship. The constant efforts to restrain the shouters' exuberance included the rigid exclusion of choruses and spiritual songs from the official Methodist hymn book.[18] Hudson states that the ring shout was preserved only among the "isolated hill people of southern Appalachia" and the "Negroes,"[19] but otherwise he exhibits a total lack of interest in the specific contributions African Americans made to the development of shouting in early American Methodism. He traces the Old World origins of shouting to the Irish and not to the Africans.

Ann Taves's insightful "Knowing through the Body: Dissociative Religious Experience in the African- and British-American Methodist Traditions" provides a more balanced description of the influence of Africans and Europeans on the progressive ritualization of the shout and other dissociative experiences in American Protestant Christianity. She traced the African impact on ecstatic religious expression in American revivalism to the eighteenth century, quoting the August 1789 report of the Methodist preacher James Meacham to the effect that "the dear black people was filled with the power & spirit of God and began with a great Shout to give Glory to God."[20] Much of the Africanization of American Protestant religious dance took place under at least nominally Methodist auspices, the earliest account being that of John Watson, a Methodist preacher, who in 1819 observed dancing "very greatly like the Indian dances" in the "blacks' quarter" at the Philadelphia Annual Conference.[21] Taves related shouting to African culture and to the experience of slavery. The greater openness of traditional African culture to dissociative religious experience and the crisis of enslavement led African Americans to elaborate and institutionalize dissociative experience at the heart of their worship life in ways that European Americans typically did not.[22]

Taves offered two generalizations contrasting African and European approaches to the sacred, namely, (1) that music and movement are largely divorced in the European approach and integrally related in the African and (2) that power is normally linked with equilibrium (balance, control) in the European context and with movement (transformation, change) in the African context.[23] This suggests that Europeans may prefer static forms of spiritual empowerment, whereas Africans prefer ecstatic or dynamic forms. Moreover, Taves's detailed discussion of dissociative religion as a problem in the academic study of religion is a further indication of the exilic status of ecstatic liturgical dances and the people who perform them.

The Chant of Affirmation: "Yes, Lord!"

Another of the salient marks of exilic worship in the Sanctified tradition is the chant of affirmation, originated by Bishop C. H. Mason in the early days of the Church of God in Christ. Pearl Williams-Jones observed that the chant of "Yes, Lord" typically follows and brings closure to the ritual shout:

> Shouts may conclude informally through the intuitive consent and feeling of tensions released by the collective body, or may give way to a chant in slow tempo such as, "Yes, Lord" which is an unmetered chant originated in the early days of the Church of God in Christ. . . . Bishop Charles Harrison Mason was heard to enchant, "Yes, Lord, Yes, Lord, Yes to your word. Yes to your will. Yes to your way." The congregation chants in heterophany.[24]

The chant of affirmation has already been cited in the excerpt from Shopshire's narrative in which the bishop exhorts the worshipers to say "yes" and dance. Murphy also describes the chant of affirmation in his narrative of worship in the Church of God in Christ:

> Mother Hall chants the Church of God in Christ national anthem, "Yes Lord." In a sure, husky voice she asks the congregation to affirm the wonders of creation, the saving deeds of Jesus, and the power of the spirit. With each pause the congregation affirms "Yes, Lord." As the enthusiasm grows, more and more people shout "Yes" and "Yes, Lord" as they feel moved. One woman comes out into the aisle to spin about with back bent, feet pumping in place, and hands raised high, fingers spread. "Oh Yes, Lord!"[25]

The chant of affirmation is sung with attendant gestures of submission such as lifting up holy hands, shouting, and cries of "Hallelujah," "Glory," "Thank you, Jesus," or simply "Yes." Ethically speaking, there is a dialectic inherent in these signs of surrender; to say yes to God is to become empowered to say no to the world, especially to the powers of evil and deception that would hinder the believer from having peace with God. Thus, the worshipers are exhorted repeatedly to drop their inhibitions and release themselves to follow the lead of the Spirit in worship. This release requires the full assent of the individuals. In this light, the inhibiting factor is ultimately sin or self-centeredness or even class consciousness. To say or sing "Yes" to God is to affirm God's acceptance of the sacrifices of praise and to signal divine approval of the saints' worship in all its culturally aesthetic concreteness and particularity. The chant of affirmation is not exclusively sung in the Church of God in Christ. The same chant or some version of it can be observed in other Sanctified churches. One alternative form of the chant of affirmation is the gospel chorus "I'll Say Yes, Lord, Yes" (to your will and to your way). This may well be the one aspect of exilic worship that is least comprehensible to those outside the tradition.

Liturgical Attendants Wearing White

A visually striking feature of Sanctified church worship is the performance of specialized liturgical roles by women, such as deaconesses, ushers, attendants, and nurses. These uniformed attendants almost invariably wear white, a color that signifies purity and consecration. Most ushers and nurses are women, and most preachers are men, but there is sufficient flexibility in fulfilling these roles to allow men to serve as ushers or nurses,[26] and women as preachers, even in the churches that do not ordain women. Even so, the women more consistently wear white when performing the liturgical roles of deaconess, usher, nurse, and preacher. White is almost always worn by deaconesses, especially on those Sundays when they are responsible for preparing the Communion table, and is typically worn by women preachers. Candidates for baptism by immersion usually wear white. Women's Day, an annual observance first instituted in the churches of the National Baptist Convention by Nannie Helen Burroughs in 1907 "to raise women, not money," is observed today in almost all black churches.[27] It is the one Sunday in the year when all the women worshipers are expected to wear white. In no way is the wearing of white an indication of a preference for white culture or assent to the biased color symbolism of a racist society. In ethical perspective, it seems to be more indicative of a desire to surrender all marks of personal style and distinctiveness to become totally identified with the worshiping community and its God, and white is the one color that makes it possible to achieve complete aesthetic uniformity.

Inclusion of Welcome and Announcements as Liturgy

At first glance, reading announcements and welcoming visitors appear to be mundane matters of marginal importance to the real business of worship. However, the announcements are an important sign of the role of the black church as a critical forum for sharing information concerning both religious and secular events, and they may bear an implicit moral approval or ecclesiastical endorsement of outside activities. The announcements provide a means for the church to proclaim that particular occurrences and accomplishments—personal, social, cultural, and political—merit the attention and affirmation of the community.

Similarly, welcoming visitors may seem to be unimportant, but the practice may have understated ethical significance as an opportunity to reaffirm and underline the open door policy of the church in the course of the liturgy. In other words, in a nation whose history includes the racist exclusion of African Americans from virtually all social institutions, including Christian churches, welcoming and recognizing each visitor by name can be a countercultural act that affirms the oneness of all humanity before God. During the aftermath of slavery, the public announcement of one's name, hometown, home church, and pastor in black church services was not only an important means of establishing personal and social identity but also a vital networking function, as Cheryl Townsend Gilkes has noted, for the reunion of individuals and families who had been sold away from each other as slaves. C. Eric Lincoln has addressed the historical significance of the practice of inviting visitors to identify themselves in worship, with reference to the black churches in general:

> The time was when the personal dignity of the Black individual was communicated almost entirely through his church affiliation. To be able to say that "I belong to Mt. Nebo Baptist" or "We go to Mason's Chapel Methodist" was the accepted way of establishing identity and status when there were few other criteria by means of which a sense of self or a communication of place could be projected.[28]

He further argued that the social identity and self-perception of black people are still refracted through the prism of religious identity. The welcome to visitors in worship offers the individual an opportunity to give voice to this identity in the community of faith. The gathering of exiles for worship always seems to call for specialized rituals of welcome, homecoming, and communication of information.

Ecstasy and Epistemology: "Having Church"

This analysis of the static structures and ecstatic expression of this worship tradition reveals a rich epistemology of sanctified religion, where song, speech, and dance all represent ways of knowing God and verifying spiritual revelation. The tradition thrives upon the integration of aesthetics (cultural authenticity), ethics (implementation of Christian norms), and epistemology (ways of knowing) in its characteristic verbal and bodily articulations of praise. Worship practices and experiences are continually interrogated with reference to specific aesthetic expectations and ethical standards. When a soloist or instrumentalist has pushed the congregation to the brink of ecstasy with an inspired performance, when the preacher has brought the sermon to a dramatic climax, and when the gatekeepers of pulpit and pew usher the people through the experience of the shout, it is understood as the "witness of the Spirit," the much sought manifestation of the Holy Spirit. After all, the underlying ethical and theological context of sanctified worship is the corporate testimony of being "saved, sanctified, and filled with the Holy Ghost."

Often the saints indicate that the ultimate purposes of worship have been fulfilled in a particular worship service by simply declaring, "We had church!" "Having church" in the Sanctified tradition requires three indispensable components: a ministering actor or artist, a responsive audience, and a spiritual anointing of divine manifestation or presence. It is not uncommon to discover these three factors in cultural and communal contexts outside worship, such as concerts, plays, lectures, political events, and social gatherings, and anyone who possesses spiritual knowledge is expected to give testimony that "we had church" when these components emerge in secular settings. Indeed, black folk have a reputation for "having church" in the streets, on the concert stage, or wherever there is ministering artist, a responsive audience and a discernible anointing of divine presence. The most widely recognized expression of the saints "having church" is gospel music, the distinct sounds and rhythms of the exilic liturgy that are the subject of the next chapter.

4

Singing the Lord's Song in a Strange Land: Gospel Music and Popular Culture in the United States

Gospel music is an American music form developed and nurtured by the saints at worship. It is both the product and by-product of "having church" in that it serves the primary purpose of supplying appropriate sounds, lyrics, and rhythms for use in worship and provides recordings and performances for broader consumption outside the church as a by-product. Gospel music provides the accompaniment for the shout or holy dance. The liturgical chant of affirmation, "Yes, Lord," is a part of the genre. As exilic liturgy, gospel music is the characteristic twentieth-century African American Christian response to the Psalmist's query: "How shall we sing the Lord's song [have church] in a strange land [in exile]? (Psalm 137:4, kjv).

In their landmark sociological study of the black churches, C. Eric Lincoln and Lawrence H. Mamiya found that the vast majority of black churches (96.9 percent) approved of the use of some form of gospel music and only 1.5 percent did not approve. Even more churches approved the use of Negro spirituals (97.1 percent). Their findings also show that 74.2 percent of black churches disapproved the use of jazz, blues, and other unspecified types of black music. In order to meet the need for diverse musical styles in worship, the average black church has three (actually 2.89) choirs, according to Lincoln and Mamiya. Another way of interpreting the same data is to say that 79.7 percent of the churches have more than one choir.[1] Although not specified in these data, in actuality the pattern of having multiple choirs was established with the advent of gospel music in the 1930s, and to this day, almost every black church has a "gospel" choir or chorus. Strangely enough, more attention is devoted in the study to explaining the resistance to gospel music on the part of the minuscule minority than is given to accounting for the appeal of the music to the overwhelming majority. They concluded that one of the current trends in black church music is to "find appropriate ways of incorporating musical styles like jazz and blues and even modern dance into worship settings," yet it is obvious that gospel music includes these and other elements of black cultural production in its very core.[2] Their explanation of the resistance to gospel music that exists among prominent segments within elite black Baptist and Methodist churches and among other "traditionalists" includes several factors: the commercialization that presupposes secularization of the music, making its metaphors and embellishments unacceptable for worship; reliance upon personal theology without consideration of any official theological canon; continuity with the evangelical tradition of Charles A. Tindley and Thomas A. Dorsey in what the authors call a "musical retreat from what is happening in the black community rather than a response to it"; and denominational bias based on the "strong Pentecostal identification" of gospel music.[3] This Pentecostal identification, however, is the appropriate point of departure for any attempt to appreciate the sacred context, meaning, and function of this music.

Outside the church, gospel music has permeated American popular culture through the media of radio, theater, cinema, and television. Radio is an especially powerful medium because of its accessibility to large listening audiences in both urban and rural areas, including those who cannot or choose not to attend churches and concerts to experience gospel performances firsthand. Portia Maultsby has observed that radio became the major source of entertainment in the United States in the 1920s, when Sanctified churches began doing live Sunday morning broadcasts that included gospel music. By the 1930s, gospel quartets and choirs were doing live performances.[4] The use of this music on the radio was not without controversy, however. Although he was not a part of the Sanctified church tradition, Thomas A. Dorsey felt strongly about the necessity of using this music in its proper context. He stated his objections to the jazz dance bands playing Negro spirituals on the radio in a 1941 letter published in the black newspaper, the *Chicago Defender:*

> Spirituals should be used only in the church. . . . It not only cheapens the songs for the bands to jazz them, but desecrates and invalidates a thing that is true to our heritage and authentic of our Race. . . . I have written more than three hundred gospel songs and spirituals. I do not object to them being used on the air, but they must not be desecrated or used for dance purposes.[5]

The popularity of gospel radio programming has increased steadily since those years, and, outside the churches, radio remains the chief medium for the promotion of gospel recordings and concerts. Maultsby is of the opinion that gospel music "took a back seat to the hegemonic programming of Black popular music" in the 1970s. It resurfaced as a viable commercial product in the 1980s, giving rise to several full-time gospel-formatted stations, which, in turn, generated an expanding consumer market for gospel recordings.[6]

Black-oriented FM stations in urban areas often reserve Sunday mornings for broadcasting gospel music, especially those recordings with a contemporary sound that have appeal to rhythm and blues listeners. Radio remains the medium of choice for most independent black religious broadcasters and has been the key to commercial success for many gospel recording artists.

The theater has become a significant setting for exposing the artistry of gospel musicians to a wider public. The first gospel musical was the 1959 production *Portraits in Bronze,* starring Bessie Griffin, an adaptation of Langston Hughes's *Sweet Flypaper of Life.* The first Broadway musical featuring gospel music was *Black Nativity,* which showcased the talents of Alex Bradford and Marion Williams. Bradford also played in Vinette Carroll's *Your Arms Too Short to Box with God,* in which in one memorable scene Jesus, newly resurrected from the dead, does a liturgical dance. A steady stream of gospel musicals has been produced since the 1960s, including *Mama, I Want to Sing; Sing, Mahalia, Sing;* and *The First Lady,* starring Vickie Winans. A unique theatrical production of the 1980s was *The Gospel at Colonus,* a modern adaptation of an ancient Greek tragedy written in the idiom of gospel music. This production, which toured the United States, released a soundtrack, and was broadcast on the public television network, used the black Pentecostal liturgy as a vehicle for expression of eternal, universal themes. Written by white composers Bob Telson and Lee Breuer, who also performed as accompanists, *The Gospel at Colonus* elevated the gospel music idiom in theater to unprecedented heights of artistic excellence by featuring exemplary performances of choral, quartet, solo, and a cappella gospel singing.

Two kinds of films have featured gospel music as a central subject or vehicle: feature-length Hollywood productions and independently produced documentaries. In 1929 King Vidor highlighted black sacred music in one of the first sound films ever produced, *Hallelujah* (1929). Relying upon the musical talents of Quincy Jones and Andrae Crouch, Steven Spielberg used gospel music to great effect in his cinematic adaptation of Alice Walker's womanist novel *The Color Purple* (1986). Near the film's end, as the people follow the wayward but repentant blues singer Shug from the juke joint and into the church, the "Yes, Lord" chant serves as a prelude to the shout "God Is Trying to Tell You Something." *Leap of Faith* (1992) features gospel performances by Albertina Walker, Edwin Hawkins, and Patti LaBelle in a plot involving an exploitative white evangelist. The best and most comprehensive documentary film on gospel music is *Say Amen Somebody* (1982), featuring interviews and performances by Thomas A. Dorsey, Sallie Martin, Willie Mae Ford Smith, the Barrett Sisters, and the O'Neal Twins.

Beginning with *Good Times* in the 1980s, gospel music has frequently been used or referenced in the theme music of television situation comedies oriented toward black audiences. Although its lead role portrayed a shrewd black deacon in an ambivalent portrayal of black religion, the show *Amen!* opened with a rousing rendition of the Andrae Crouch composition "Shine on Me." The major television networks have given increased exposure to gospel music and artists in special productions aired during the Christmas holidays, in the music awards programs, and in advertisements. The 1994 National Football League Super Bowl featured a gospel rendition of the national anthem by Natalie Cole and the combined choirs of several black colleges and universities in Atlanta. The production and distribution of gospel videos represents an area for potential growth in the creative presentation of gospel performances. In general, however, the theatrical, cinematic, and televised portrayals of gospel music have exploited its entertainment value and have tended to trivialize, ridicule, or denigrate black religion.[7]

The historical development of gospel music has been the subject of books, dissertations, and articles.[8] One approach to understanding how this music came into being and continues to change and develop is to examine several key trajectories in the formation of the genre. A trajectory is a path or track followed by an object in motion. Pearl Williams-Jones, a professor and performer of sacred music, defined gospel music in two ways: as a synthesis of West African and African American music, dance, poetry, and drama and as a body of urban contemporary black religious music of rural folk origins that celebrates the Christian experience of salvation and hope.[9] On the ground of this understanding of gospel music as a synthetic art form, it is possible to trace and identify several separate trajectories representing traditions whose merger into a single liturgical form constitutes the artistic foundation and ongoing creativity of gospel music: European Protestant hymnody, Negro spirituals, blues, jazz, rhythm and blues, rap, and classical music. Illustrations of the current impact and importance of each trajectory are cited from contemporary gospel recordings and worship practices.

European Protestant Hymnody

The first trajectory is European Protestant hymnody and the gospel and inspirational music that subsequently evolved in the white evangelical culture of the United States. Michael Harris traced the phrase "gospel music" as far back as nineteenth-century England, where Dwight L. Moody's music director, Ira Sankey, claimed to have witnessed the origination of the phrase "to sing the gospel" in 1873.[10] The songs of eighteenth-century British hymnwriters such as Isaac Watts, Charles Wesley, and John Newton were embraced by African American converts to evangelical Christianity and are included in the sacred repertoire of many of the contemporary Sanctified congregations. White American hymnwriters of the nineteenth and twentieth centuries have also contributed to that repertoire. For example, the music of Sankey, the "father of the gospel song," who helped to popularize gospel singing through his evangelistic work with Moody and his publication of gospel songbooks, and of Fanny Crosby, who perhaps wrote more popular gospel songs than anyone else, is still being performed and recorded by gospel artists. The lyrics for "Amazing Grace" were written by former slave ship captain John Newton, but the melody was first known as a plantation tune entitled "Loving Lambs."[11] Perhaps this peculiar pedigree accounts for its abiding importance as a standard gospel hymn; it was sung by gospel pioneer Mahalia Jackson, and in 1972 Aretha Franklin released *Amazing Grace,* a double album of her own arrangements, lead vocals, and piano accompaniment, performed with James Cleveland and the Southern California Choir.

The white Protestant gospel songs and hymns influenced the development of black gospel music in at least two ways. First, because they were accepted and sung by black participants in revivals and evangelistic churches, they offered lyrics and messages that could be freely borrowed and incorporated into new black gospel compositions. Second, they were given new arrangements and embellishments after the advent of the distinctive gospel musical style and were effectively absorbed into the repertoire of black sacred music.

White and black artists continue the pattern of borrowing and embellishment in the realm of contemporary sacred music, as exemplified by the collaborative efforts of Richard Smallwood and Bill and Gloria Gaither, who wrote the immensely popular "Center of My Joy" together, and the duets of Sandi Patty and Larnelle Harris, such as "More Than Wonderful" and "I've Just Seen Jesus." Not coincidentally, Patty and the Gaithers are from the interracial Church of God (Anderson, Indiana). Patty's music is widely played by black gospel radio stations, and her signature solo, "We Shall Behold Him," was "embellished" in a popular recording by Vickie Winans. The Gaithers are the authors of "He Touched Me," which has become incorporated into the repertoire of black sacred music as a classic. On their first two recordings, the a cappella male sextet Take 6 included innovative arrangements of Ralph Carmichael's "A Quiet Place" and "The Savior Is Waiting." Even country music, the major influence in white inspirational music, has made its mark on the black gospel tradition. Shirley Caesar's popular single "No Charge" was released in two versions, the first side "gospel" and the flip side "country," the main difference being that the country version omitted her preaching interludes. The country sound was also dominant in the biggest gospel "hit" of the Reagan era, "(I'm Coming Up) the Rough Side of the Mountain," a duet by F. C. Barnes and Janice Brown, who, like Caesar, are North Carolinians and pastors of a Sanctified congregation.[12] As further evidence of the importance of this trajectory, the 1969 recording that marked the entry of gospel music into American popular culture as a marketable art form, Edwin Hawkins's "O Happy Day," was a Pentecostal rendition of an eighteenth-century British hymn.[13]

Negro Spirituals

A second trajectory in the development of black gospel music is represented by the Negro spirituals that emerged from the crucible of slavery. Via this trajectory, the African influences were perhaps most strongly impressed on the living tradition of black sacred music. Gospel music, like the Negro spirituals, is a product of African Christian liturgy; its Africanness is manifested primarily in performance style, and its Christian message issues forth in the vernacular expressions of worshipers and performers who are confessing the good news of Jesus Christ against a backdrop of deprivation and suffering. The "original" Negro spirituals were songs composed and performed in the context of worship, typically in the "hush arbors" and clandestine gatherings of the so-called invisible institution. Historian Albert J. Raboteau has discussed the process by which the gospel message first became incorporated into the indigenous music of the African American slaves: "Unable to read the Bible for themselves and skeptical of their masters' interpretation of it, most slaves learned the message of the Christian Gospel and translated it into songs in terms of their own experience."[14] The full communal and liturgical setting in which the spirituals were sung was the prayer or praise meeting. Like twentieth-century gospel music, these songs were communal songs, "heard, felt, sung and often danced with hand-clapping, foot-stamping, head-shaking excitement."[15]

Raboteau's emphasis upon the liturgical performance setting is critical because in the present century the Negro spirituals are known mainly in their concert form or by their lyrics as reformulated and regularized by John W. Work and others. Work, a Fisk University professor who led the Fisk Jubilee Singers on concert tours throughout the United States from 1900 to 1916, sought to establish a concert repertoire of Negro spirituals suitable for white audiences.[16] Many scholars, especially black theologians, have written exhaustive analyses of these songs based on the words alone, without necessarily giving adequate attention to the impact of context and performance style. Michael Harris, by contrast, has crafted an exemplary analysis of the formative role played by the performed spiritual and its "bush arbor" context (as replicated in the urban black churches outside the South) in the emergence of the gospel music genre under the artistic leadership of Thomas A. Dorsey, Sallie Martin, and Mahalia Jackson. While the gospel artists were negotiating the acceptance of their music in black churches that emulated white Protestant music and worship, he explained, the "bush arbor" was becoming institutionalized in the Sunday sanctuary worship of the Holiness churches and in the midweek and Friday testimony meetings held in the basements of those same black denominational churches.[17] Ultimately, the Negro spiritual could not retain its full aesthetic and liturgical integrity in the course of the many transformations that occurred over time in African American Christian exilic consciousness and assimilationist attitudes. Instead, the spiritual

> [T]ook on another life and identity with the "jubilee" singing tradition and with the Works, Burleighs, and Detts who sought to "dress it up" as an appropriate song form for the "new" Negro . . . with its powerful creative texts now frozen and its call-and-response structure replaced by chordal, Anglophonic harmony. But even in this disguise it served nobly in the accommodation of the African American presence in the strange land of white mainline Protestantism.[18]

Williams-Jones was similarly critical of the cultural inauthenticity of "arranged" spirituals, which she set in contrast to the aesthetic integrity and improvisational genius of the black gospel genre: "Black gospel music has not consciously sought the assimilation of European religious music practices or materials into its genre. If this has occurred, the materials have been improvisationally recreated to conform to black aesthetic requirements of performance."[19]

Notwithstanding the compromises in form and performance that were imposed upon the spirituals in the interest of textual preservation and mainstream acceptance, this trajectory remains a vital force in gospel music. Some contemporary gospel recording artists are forging robust new interpretations of the Negro spirituals that may assist the black churches in rescuing this genre from its alleged cultural and liturgical captivity. One noteworthy example of this potential trend is the 1994 release by Sounds of Blackness, *Africa to America: The Journey of the Drum,* which uses Gary Hines's abbreviated arrangements of the Negro spirituals "Hold On" and "Ah Been 'Buked" to frame the presentation of several other musical styles representing the African American cultural and spiritual heritage.

The Blues

The blues is an indigenous African American folk genre that expresses the melodies and rhythms of lament in a clearly delineated set of harmonic progressions. The formative context for blues performances is entertainment instead of worship—especially the bars, dance halls, clubs, and bordellos of the urban and rural South. Thomas A. Dorsey was an accomplished blues pianist who initiated the blues trajectory into contemporary black sacred music as he worked with the black churches and choirs of Chicago in the first part of this century. Harris cited Dorsey's own description of the distinctive mood and context and of the blues:

> Some have asked what is the blues. It would be hard to explain to anyone who has never had a love craving, or had someone they loved dearly to forsake them for another, a wounded heart, a troubled mind, a longing for someone you do not have with you. . . . Blues would sound better late at night when the lights were low, so low you couldn't recognize a person ten feet away, when the smoke was so thick you could put a hand full of it into your pocket. The joint might smell like tired sweat, bootleg booze, Piedmont cigarettes and Hoyttes Cologne. . . . The piano player is bending so low over the eighty-eight keys, you would look for him in time to swallow the whole instrument. He is a king of the night and the ivories speak a language that everyone can understand.[20]

It is this capacity to capture and articulate emotion that Dorsey eventually imported into the sacred context of worship.

The precipitating event for Dorsey's cathartic merger of a blues style and text into a sacred form, "giving the gospel a blues voice," was the sudden death of his young wife and infant son. In the aftermath of this tragedy, Dorsey wrote his signature composition, "Take My Hand, Precious Lord," as he allowed himself to wail, to get "lowdown," to purge—rather than just soothe—his grief.[21] Harris explained that this "marriage" of Dorsey's musical and textual voices resolved deep conflicts in Dorsey's life as a blues musician performing sacred music, and between "old-line" (i.e., the black Protestant churches) and indigenous African American religion.

To Dorsey's great surprise, his initial performance of "Take My Hand, Precious Lord" elicited uncharacteristically ecstatic responses at the Ebenezer Baptist Church in Chicago: "The folk went wild. They went wild. They broke up the church. Folk were shouting everywhere."[22] Thus, the blues trajectory in gospel music was established when, in the midst of his own deep distress, the bluesman known as "Georgia Tom" discovered how to use the blues idiom for emotional release in the context of worship. Ray Charles, who is probably the most popular vocalist and pianist in the African American blues tradition, has collaborated with Take 6 in recording the gospel blues selection "My Friend" in *Join the Band* (1994). Another gospel interpretation of the blues, "Livin' the Blues," is recorded in *Africa to America* (1994) by the Sounds of Blackness.

Jazz

As a musical style, jazz is known by its strong rhythmic understructure, solo and ensemble improvisations on basic tunes and chord patterns, and highly sophisticated harmonic idiom.[23] The original meaning of the word *jazz* is "sex"; it is derived from the Creole patois *jass,* a sexual term applied to the Congo dances in New Orleans. Apparently the term reflects the sensuality of the music, its mood, and its context. Leonard Barrett has commented that dance was the most prominent element in the behavior of African slaves; many slave dances concerned fertility rites and emphasized pelvic movements.[24] He further explained how the prominence of the drum in the slave culture in New Orleans, unlike other regions, where drums were forbidden or suppressed, was a factor in the rise of jazz: "Here jazz had its origin, in New Orleans and not in Boston or Philadelphia, solely because the Africans had access to the drum."[25] In his description of slave religion in New Orleans in the 1850s, Sterling Stuckey has brought attention to the city's reputation for Congo square dances and public expressions of Africanity.[26] Stuckey believes that the sacred dance of the slaves, the ring shout, and secret societies helped to form the context in which jazz music was created in Louisiana.[27] With attention to developments in New Orleans at the beginning of the twentieth century, Harvey Cox has cited several additional factors that "coalesced" to give jazz its start "in the steamy back streets of New Orleans," including the highly complex counterrhythms of the African slaves, the availability of secondhand military band instruments after the Spanish-American War ended in 1898; the ecstatic hymnody of

Southern revivalism, the socially coercive effects of the state legislature's redefinition of the legal meaning of race, and the legalization of prostitution (which created new employment opportunities for musicians in bars, restaurants, and hotels).[28]

For these and perhaps many other reasons, jazz came to prominence as a genre with secular and sacred roots. Maultsby declared that the "bluesy," "jazzy," and "rockin' " sounds from instruments such as the tambourine, drums, piano, guitar, horns, and organ brought a "secular dimension" to black religious music.[29] Although these musical instruments were banned by some churches as tools of the devil, the Holiness churches were the first to employ them in the service of God. Stuckey claimed that the most famous New Orleans tune of all, "When the Saints Go Marching In," was originally a Sanctified shout, celebrating the "saints" who have followed Jesus all the way.[30] Cheryl Townsend Gilkes alluded to the same tune in her discussion of the emergence of the Sanctified church tradition as an "assertive black cultural revolution which fueled the sacred and secular popular culture of the American South and eventually the entire culture (e.g., Oh when the Saints go marching in)."[31]

Duke Ellington is the jazz artist best known for exploring sacred themes in his compositions and performances. The jazz trajectory in gospel music took a significant turn in the 1950s with the advent of the bebop era, which produced what Cornel West has identified as the "Africanization" of jazz "with the accent on contrasting polyrhythms, the deemphasis of melody, and the increased vocalization of the saxophone."[32] Many of the jazz musicians of this era who grew up near the Sanctified churches were familiar with its music, and Stuckey claimed to have witnessed Theolonius Monk doing a ring shout in the middle of a jazz performance, "his feet beating out intricate figures before he returned to the piano and joined his combo in playing music as advanced as any of his era."[33] With reference to two of John Coltrane's 1960s recordings, "A Love Supreme" and "Meditation," Cox declared that "no one embodies better than Coltrane that strange kinship between Pentecostal incantation and the spiritual lineage of jazz."[34]

The New Orleans connection in the jazz-gospel trajectory is manifesting itself afresh in the musical ideas of Wynton Marsalis, a New Orleans native who did not grow up in the church himself but has been influenced by musicians in his group who grew up playing church music. He describes his motivation for writing and recording *In This House:*

> Almost everyone in the band grew up playing church music and what truly spurred my desire to write this music was the many hymns and shouts that they sing on the bus as we travel, at sound checks before concerts, and after

meals. With the demise of a viable blues tradition in popular music, most of the younger jazz musicians learn the expression necessary to play music either in church or from someone close to them. . . . Listening to all of them made me want to put that feeling in a long piece and reassert out here the power that underlies jazz by constructing a composition based on the communal complexity of its spiritual sources.[35]

This secular jazz composer drew upon the mutual interplay of formative influences manifested in the artistry of church-bred jazz musicians. Moreover, he used the idiom of jazz to merge the blues with the music of the black church: "My intention was to reconcile the secular nature of blues expression with the spiritual nature of its sources."[36] Marion Williams's vocal performance of the selection "Prayer" echoed the chant of affirmation and "Yes, Lord" known in the black Pentecostal churches.[37] Jazz critic Reuben Jackson has commended Marsalis for his willingness to "risk incorporating the sensuality so deeply ingrained in African American worship." In his opinion, Marsalis was inspired to do his best solo recording ever in "Sermon," the climactic selection that most clearly expresses the characteristic rhythms and vocalizations of the Pentecostal shout.[38]

Rhythm and Blues: Soul Music

The term "rhythm and blues" was first used in 1949 by *Billboard* editor Jerry Wexler in an effort to create an updated designation for black records featuring dance rhythms and blues confessionals. Rhythm and blues is a combination of gospel-style singing with romantic subjects.[39] This trajectory in gospel music is the one most readily identified and widely recognized, mainly because of the public prominence and phenomenal commercial successes of rhythm and blues artists with church "backgrounds." In his article on ritual structures in African American music, anthropologist Morton Marks argued that gospel is the repository of African-based performance rules in the United States. He saw soul singers such as James Brown, Aretha Franklin, and Wilson Pickett as threshold figures whose performances turn into trance rituals almost indistinguishable from "shouts" in church.[40] The proof Marks offered in support of this argument is the widespread practice of "covering." A cover record is a copy of an original recording performed by another artist in a style thought to be more appropriate for the mainstream market. Usually black artists recording for independent labels were covered by white artists signed to major labels; in the 1950s, covers were used to "capitalize on the growing popularity of rhythm and blues among white listeners."[41] Marks has observed that when gospel and soul groups "cover" each others' recordings, trance-associated features such as preaching, hyperventilation, and paradoxical communication are preserved.[42] Thus, a strategy originally designed to exploit the artistic genius of black musicians eventually became the vehicle for introducing the ecstasy of Sanctified spirit possession to the world through the medium of gospel and rhythm and blues.

The "soul" music of the 1960s employed the sounds of gospel and rhythm and blues to signal a cultural return to the African past; it gave voice to African exiles' experience of pain and suffering, of love and hope, in America. It is not entirely clear which came first, the music or the black power revolution, but both mirrored the sentiments of black urban dwellers primed for social change. What is clear is that the primary context of this new genre was not the church but the house party. West has aptly described the emergence of soul music in this light:

> Soul music is more than either secularized gospel or funkified jazz. Rather, it is a particular Africanization of Afro-American music with intent to appeal to the black masses, especially geared to the black ritual of attending parties and dances. Soul music is the populist application of bebop's aim: racial self-conscious assertion among black people in light of their rich musical heritage. The two major artists of soul music—James Brown and Aretha Franklin—bridge the major poles in the Afro-American experience by appealing to agrarian and urban black folk, the underclass and working class, religious and secular men and women.[43]

In West's judgment, the greatest album in the history of black popular music is *What's Going On* (1971), produced by Marvin Gaye, a singer who emerged from the Pentecostal church.[44] Significantly, a year later, Aretha Franklin covered a selection from Gaye's signature album, "Wholly Holy," in her *Amazing Grace* recording, which, together with Edwin Hawkins's *O Happy Day*, firmly established gospel music in the marketplace of American popular culture.

The dynamic tension between dance music and church music is still being played out in the burgeoning gospel music industry, as evidenced in *Africa to America* by Sounds of Blackness, whose spirituals and gospel ballads were inserted among the inspirational funk tunes produced by Jimmy Jam and Terry Lewis. As one music critic has commented, this recording has transformed funk beats and pop hooks "from erotic scenarios into a broad tapestry of spiritual and cultural concerns. . . . The result is an album that blends the "mass choir" phenomenon of the gospel world, the radio-ready funk of the pop world and a hopefulness rarely heard in these bleak days."[45]

The rhythm and blues trajectory has brought innovative sounds and increased marketability to gospel music and received in exchange a message of hope and permission to invoke the power of the Spirit outside the church. In this light, it can be judged that the greatest gift of the Pentecostal churches to American popular music is the rhythm and blues of Marvin Gaye and not the rock and roll of Elvis Presley, who emerged from white Pentecostalism with an uncanny ability to cover black music and adopt black performance styles. Unfortunately, both came to tragic, premature deaths as the consequence of self-destructive and dysfunctional behaviors contradictory to the life-affirming message of the gospel.

Gospel Rap

The rap trajectory in gospel music is a new development rooted in an old tradition. The dynamics of the spoken word are incorporated into the characteristic vocalizations of gospel music by means of three major speech forms of Sanctified church worship: prayer, preaching, and testimony. In other words, gospel music speaks the words of prayer, of preaching, and of testimony in forms familiar to black worshipers. Williams-Jones has described the relationship between prayer and music in the gospel tradition:

> Song is born on the wings of prayer. Individual or collective improvisational prayers contrast markedly with the liturgical, formal prayers of white churches in the European worship tradition. Prayers may be moaned, sung, chanted. Again, patterns of call and response are not uncommon in collective praying. It is almost impossible to say when a prayer ends and a song commences as music and prayer intertwine.[46]

Her understanding of the aesthetic ideals of gospel music leads to the conclusion that the gospel singer is "the lyrical extension of the rhythmically rhetorical style of the preacher."[47]

The sacred speech that Marks referred to as "song-sermons" in his analysis of the performance rules of African American gospel music, " 'You Can't Sing Unless You're Saved': Reliving the Call in Gospel Music," would be more accurately described as "song-testimonies." His analysis focused on songs by the Pilgrim Jubilee Singers and the Swan Silvertones, whose lead vocalists "relive in musical form the moment of their conversion," which is precisely the function of testimony in the tradition. He explained that the special language of converts and initiated performers in the black Protestant churches, spoken in the form of black preaching and testimony, exemplifies the "cult languages" found in other African-derived religions. This special language does not necessarily include glossolalia; Marks pointed out that the ecstatic utterance of tongues that has priority in the Pentecostal and Apostolic churches is not usually a major factor in gospel music.[48] What is important in the music is the testimony of being saved, an indication of the permanent relationship with the deity underlying the role of the preacher-singer. The critical aspect of testimony, as opposed to preaching, is its subjectivity; testimony conveys one's story of divine encounter in strictly personal terms, with or without a biblical or cultural text. Moreover, one does not need special credentials or sanction to testify, so that even in churches where women are not ordained as preachers, they are authorized, even required, to give spoken testimony of conversion and sanctification.

A striking example of a gospel song-sermon is Aretha Franklin's spoken interlude in "Mary, Don't You Weep," in which she used to great effect the preacher's technique to relate the story of how Jesus raised Lazarus from the dead. Her performance received the full endorsement of her father, the renowned Baptist preacher the Reverend C. L. Franklin, whose remarks appear elsewhere on the *Amazing Grace* album. Noteworthy also is the gospel song-testimony of Shirley Caesar. Her recording "Hold My Mule" featured her spoken narrative about a man who was forbidden to shout in a "dignified" church. When the deacons come to his farm and interrupt his plowing in order to reprimand him further, he begins to testify about the goodness of the Lord and asks for someone to hold his mule so he can shout on the spot. Both of these examples represent precursors to the rap forms now being implemented in gospel music. It is somewhat ironic that in gospel music women have "come to voice" in rhetorical forms often seen as exclusively belonging to men. As early as 1927, in *God's Trombones,* James Weldon Johnson's classic little book of poetry inspired by black folk preaching, it is noted that the spoken prayers of the black church, sometimes led by a woman, "were products hardly less remarkable than the sermons."[49] Likewise, women have played a formative role in the emergence of rap music as a secular art form. West traced the modern origins of rap music to Sylvia Robinson, a soul songwriter who recorded "Rapper's Delight" by the Sugarhill Gang.[50]

Secular rap music combines the two major "organic artistic traditions" in black America—black rhetoric and black music—and "Africanizes" African American popular music with its syncopated polyrhythms, kinetic orality, and sensual energy.[51] West lamented the "lyrical hopelessness" and the radical challenge rap presents to the transcendent and oppositional roots of African American music. He argued that what is missing from rap music is the "opposition that somebody cares," as is articulated so forcefully in other African American musical forms.[52] This opposition is, of course, the core message "spoken" in gospel music via forms related to prayer, preaching, and testimony. Gospel rap borrows performance styles from the hip-hop culture and from the black churches to convey a positive message of hope and transcendence to young audiences.

Although many of the leading rap or hip-hop artists are widely condemned for conveying violent and misogynist messages, some have incorporated gospel lyrics and messages into their songs. For example, Hammer released two such recordings, "Pass Me Not" and "Pray," and Queen Latifah joined forces with Take 6 to write and record a rap selection entitled "Harmony." The increasing popularity of gospel rap artists such as DC Talk, Gospel Gangstas, and Disciples of Christ is evidence of the viability and effectiveness of this most recent trajectory in gospel music. Their performances represent a fresh articulation of the song-testimony and preaching found in other forms of gospel music:

> Christian hip-hop blends hard-core rap's edgy energy and violent imagery with traditional black gospel's fire-and-brimstone preachings. The results turn both traditional rap and gospel music upside-down. Songs rap about making "drive-bys" on Satan and whipping out a "King James switch-blade" on demons. . . . Most cry out in a tough-love appeal to listeners to put down their sinful ways.[53]

To be sure, gospel rap will have some impact upon the future development of the liturgies of the Sanctified churches. A former minister at Third Street Church of God is a skilled rapper who has written a full repertoire of songs to reach young people in youth rallies, worship services, conventions, and concerts. He regularly performs for the church's urban prayer breakfast, whose constituency is largely young black men, but when rapping in the main sanctuary he stands on the main floor and almost never in the pulpit, in deference to the saints' sensitivities concerning secular dancing and rhythms.

Classical Music

The classical trajectory in gospel music flows principally through the influence of classically trained performers of gospel music. Roberta Martin was one of the early gospel pianists who incorporated classical music styles and knowledge into the gospel genre. Pearl Williams-Jones created a widely acclaimed vocal and piano gospel arrangement of Charles Wesley's "Jesus, Lover of My Soul" set to the tune of J. S. Bach's "Jesu, Joy of Man's

Desiring"; Richard Smallwood recorded his group's version of this arrangement as a posthumous tribute to her on the 1992 release *Testimony*.[54] Smallwood offered a classically inspired piano solo performance of the beloved hymn "Great Is Thy Faithfulness" on the same release, and the prelude to his 1993 live recording at Howard University is Bach's "Jesu, Meine Freude." Another example of the classical trajectory in gospel music is *A Soulful Celebration* (1992), Quincy Jones's interpretation of Handel's *Messiah*. It is not unusual for student choirs on black college campuses to perform on the same program a repertoire inclusive of classical sacred pieces by Bach, Handel, and other European composers; Negro spirituals written and performed in the classical idiom; and contemporary gospel music.[55]

In summary, gospel music is the confluence of seven streams: white Protestant hymnody, Negro spirituals, blues, jazz, rhythm and blues, rap, and classical music. These trajectories all flow together as a testament of gratitude to the divine source of all human creativity, who is the proper subject and object of all human devotion. Gospel music is a culturally specific mode of communicating the gospel of Jesus Christ; its defining purpose, as Williams-Jones has said, is a celebration of the Christian experience of salvation and hope. Each of its separate trajectories is representative of distinctive languages, audiences, and contexts, within which that one message is being communicated.

Gospel Musicians in the Sanctified Tradition

Michael Harris, Anthony Heilbut, Morton Marks, Cornel West, and others have in their own way cited the importance of religion in the formation of artists identified with the various trajectories of black music and culture. Harris has given attention to the role of religious commitment in Thomas Dorsey's torturous transition from bluesman to gospel composer, Heilbut has provided a detailed catalog of who among gospel singers is a saint and who is not, Marks has explored the proposition "You can't sing unless you're saved" with respect to the ritual of trance in gospel performances, and West has marveled at the "troubled genius" of Marvin Gaye, whom he identified first and foremost as a Christian artist. To answer the question of the meaning of gospel music for the saints, it would seem appropriate to examine the testimony and experience of two gospel pioneers who maintained their identity and status within the Sanctified church tradition: Sallie Martin (1896–1988) and Willie Mae Ford Smith (1904–1994).

Sallie Martin was born and raised in Pittfield, Georgia. Her father died when she was young, and her family made their living by farming. She left school after the eighth grade and moved to Atlanta, where she found employment in domestic positions. She joined the Fire Baptized Holiness Church in 1916: "I'm very happy I was saved in a Holiness church, cause I got some principles from them that I'd *never* lose, *never* lose, never."[56] She moved north to Cleveland and thereafter to Chicago with her husband and son. She left her husband after he set up moonshine and gambling in their home to make money, and she worked in a hospital to support herself and her son. She met Dorsey in Chicago and, in 1932, made her solo debut with Dorsey's group at Ebenezer Baptist Church. Martin traveled with Dorsey for nine years, helped to organize gospel choruses, and was, with Dorsey and Willie Mae Ford Smith, cofounder of the National Convention of Gospel Choirs and Choruses. She is credited with many firsts. Her Sallie Martin Singers were the first female gospel group. With composer Kenneth Morris, she established the Martin and Morris Music Company in 1940, which is the oldest continuously operating black gospel music publishing firm in the United States and still the largest supplier of gospel music.[57] In 1946 she was narrator for the first gospel television broadcast, performed by the choir at St. Paul Baptist Church in Los Angeles.

Martin had a sense of call to the ministry, and, on meeting Dorsey, the two became "one in purpose before they became one in fact by virtue of their being two preachers in search of musical sermons and pulpits."[58] Not only was Martin able to find her homiletical voice by performing the expressive music being produced by Dorsey but also their commercial successes gave her an opportunity to exercise her administrative abilities and organizational skills, to the dismay of some of the male ministers. In response to these ministers's attempts to "call her in," Martin ignored them, saying that "she was doing Kingdom work and got her authority from God."[59] Speaking from the vantage point of a composer and performer of gospel music, she clearly understood the ethical connection between authenticity and effectiveness in music ministry:

> I like the old songs, I do, I really do. I think the old songs were written out of some kind of burden. The old songs wasn't some song some person sat down and said, I'm gonna get me some kind of song together to make me some money. . . . I can't sing just any song. A song that carry a message got to have some kind of absolutely *common* sense, and a whole lot of songs sing about my mother's gone and my father's gone, well we don't need all of that. They're playing on emotions, that don't mean nothin'. They're not tellin' them you got to live right, and if you're not right, you need to get right, see.[60]

Martin was able to reap and accumulate some of the material benefits of her labors (Heilbut calls her a "rich woman") and to realize her childhood dream of using her singing talent to elevate herself above and away from the abject poverty of her Southern rural upbringing. Her philanthropic projects included private scholarships, aid to less fortunate singers, and a mission in Nigeria, where a building was named for her.[61]

Mother Willie Mae Ford Smith (1904–1994) was born in Rolling Fork, Mississippi, one of fourteen children. Heilbut calls her the greatest of the " 'anointed singers,' the ones who live by the spirit and sing to save souls."[62] At the age of twelve she moved with her family to St. Louis; she quit school in the eighth grade to help run her mother's restaurant. Her father was a Baptist deacon, and she was raised as a Baptist. She formed a family quartet with her sisters, and in 1922 they gave an impressive performance for the National Baptist Convention. Beginning in the 1930s she headed the Soloists Bureau for the National Convention of Gospel Choirs and Choruses, which she cofounded with Martin and Dorsey. Mother Smith was known for her many protégés; among the most successful of them are Joe May, Myrtle Scott, and Edna Gallmon Cooke. In 1939 she joined the Church of God Apostolic:

> You know I used to love to party, to dance and to sing. Count Basie, Bessie Smith, Cab Calloway. Before I got saved, I used to have a ball. I imagine I'd have gone into blues myself. It's just, when I got saved, nothing interested me but serving God. I'm not condemning anybody, it just wasn't for me.[63]

Mother Smith then turned her talents and attention to evangelistic work. She believed that she was anointed by the Lord when singing: "He didn't give me the gifts of prophecy or healing, but he gave me a story and a song."[64] She rendered an impressive demonstration of anointed singing in the documentary *Say Amen Somebody* with her performance of "Never Turn Back." In 1988 the National Endowment for the Arts awarded her a National Heritage Fellowship in recognition of her contributions as an outstanding American folk artist. She was featured in Brian Lanker's *I Dream a World: Portraits of Black Women Who Changed America,* in which she offered a succinct summary of her ethical approach to gospel performance: "To be a gospel singer, you got to be a gospel person."[65]

Willie Mae Ford Smith and Sallie Martin were able to engage in the gospel music ministry for an extended period due to their longevity. Both women had a sense of call to ministry that they fulfilled through evangelistic performances of the music. Martin had a commonsense understanding of the ethical meaning of her music and took it seriously as the work of the kingdom. Smith's ethical concern was the charismatic exercise of music as one of several gifts of ministry that operate to help people find their way spiritually. Both favored testimonial texts of personal transformation and commitment in the lyrics of their music and made use of the spoken word to enhance the message. For these saints, the anointing of the Holy Spirit is the highest ideal and objective of a gospel performance. It is achieved by following the advice given by both singers to their young protégés: "Be original and be yourself."[66] In other words, one does not achieve or acquire the anointing of the Spirit merely by imitating others but rather by being disciplined and prepared to perform authentically with an openness to improvisation and innovation as the Spirit leads. As Mother Smith said, "You sing what fits you."[67]

Unlike most scholars of African American religion and culture, these two saints drew a definite distinction between the sacred and the secular in their music and life experience. Both incorporated "secular" styles into their singing, drawing especially from the blues tradition, but had no interest in performing blues or any other music apart from the context of worship. For the saved and sanctified performer, the selective synthesis of sacred and secular elements in the music proceeds according to the principle "in the world, but not of it." What matters here is not the sacred or secular origin of the particular style or technique being used; arguably, all sacred music borrows from corresponding secular music forms.[68] At issue is intentionality—will the performance encourage emotional release, spirit possession, shouting, conversion experiences? These things can happen when God is glorified as an aesthetic and ethical priority, the performance goal of Mother Smith: "I never could go along with those folk who say I shouted so and so many. When folk shout, I give it all to God. I carry on for his honor. If they do it out of the message, then I'm grateful. . . . If God doesn't stand up, then I just don't do."[69]

The message of the music is its most important quality. The task of the musician is to elevate that message artistically so that people are moved to hear and respond to it as a life-changing experience, rather than be entertained or impressed by the performer's gifts and virtuosity alone. Music is ministry in the sacred realm, and a vast range of secular voices, styles, and sounds are available to be borrowed in order to enhance its effectiveness and expand its audience. The purpose of this music, therefore, is spiritual formation, by any means necessary.

It is clear that Sanctified gospel singers created their own definitive answers to the question "How shall we sing the Lord's song in a strange land?" As a whole, the performers of gospel music are moved and motivated by a host of factors in what they do and have at their disposal a vast array of streams or trajectories (certainly not limited to the ones discussed here) from which to draw artistic ideas and approaches for making music. Although those who have reaped the greatest acclaim and material rewards based on this idiom have been the artists who have crossed over from worship into entertainment as a primary context for their music, the saints have tended to forgo opportunities to become stars and instead have stayed focused on the music and its message as a manner of doing ministry. They have added ethics to ecstasy. Many are the performers who can induce trance in an audience; fewer of them actually live the life they sing about in their songs.

5

Resistance, Rebellion, and Reform: The Collegiate Gospel Choir and the Black Clergy Caucus

Several scholars have illuminated aspects of the Sanctified church tradition that represent movement toward social rebellion, cultural resistance, and ecclesial reform. Historian David Daniels has contended that the worship practices developed under the leadership of C. P. Jones and C. H. Mason fostered the renewal of African American Christianity through a restructuring of slave religion. In alliance with other "progressive" black churches, the Holiness movement advocated moral, ecclesial, liturgical, and pastoral reforms.[1] Sociologist Cheryl Townsend Gilkes has portrayed the Sanctified church as a cultural resistance movement whose liturgies and structures enabled black women to resist the racist and sexist cultural assaults they experienced in the United States during the period between 1895 and World War II, the era of Jim Crow:

> Women in the Sanctified church were committed to the cause of racial uplift.
> They retained their commitment to ecstatic worship, which black Baptists
> and Methodists were rejecting. They also retained an emphasis on women's
> interests, education, professionalism, and the cultivation of a black female
> image that contradicted the dominant culture's stereotypes.[2]

Gilkes concluded that black women in the Sanctified church have maintained a dialectical tradition of protest and cooperation, driven on the one hand by their struggle against structures and patterns of subordination based on sex and, on the other by their determination to maintain unity with black men in the face of racism and discrimination.[3] Although her primary focus was the description of women's traditions in these churches, Gilkes's ultimate purpose was to discern and delineate models of cooperative and egalitarian male-female leadership within the Sanctified churches.

Walter Hollenweger has called attention to the emergence of a distinctive new unity between "prayer and politics, social action and song" in the black Pentecostal churches:

> In addition to the charisms which are known in the history of Pentecostalism, such as speaking in tongues, prophecy, religious dancing, prayer for the sick, they practice the gift of demonstrating, organizing, and publicizing as another kind of prophecy. I have known black Pentecostal churches in which these activities were explicitly mentioned in a list of gifts of the Spirit, but not as it is usually done in many political church groups in Europe where political analysis replaces prayer and song (not to speak of dancing and speaking in tongues).[4]

The relationship between Pentecostalism and black power has also been explored by Luther P. Gerlach and Virginia H. Hine in *People, Power, Change: Movements of Social Transformation*. They concluded that it is possible to identify both Pentecostalism and black power as parts of a cultural revolution.[5] James S. Tinney's doctoral dissertation also compares black political and religious movements, giving special attention to the political overtones and implications of Pentecostal beliefs. He discerned among black Pentecostals a "total obsession with power forms of one kind or another." Moreover, a modus operandi of change by revolutionary or even violent means is latent within their doctrine of expropriation of spiritual power.[6]

This impetus toward ecclesial reform, cultural resistance, and revolution or rebellion growing out of the Sanctified church tradition can be further illustrated with reference to two specific examples of organized black initiatives undertaken in response to the experience of alienation in the broader contexts of higher education and denominational life, respectively: the collegiate gospel choir movement and the black caucus movement in the Church of God (Anderson, Indiana).

Gospel Music and Black Identity on Campus

Generally speaking, gospel music is a key factor that attracts young people to the black churches. Notwithstanding their negative characterization of some of these young people as "gospel music groupies" and "part-time college-aged church goers," sociologists C. Eric Lincoln and Lawrence H. Mamiya acknowledged the cultural significance of the collegiate gospel choirs. They discovered that gospel music has found a place beside concert spirituals in the otherwise classical repertoire of black college choirs.

Moreover, separate gospel choirs were developed and eventually found wide acceptance, "although official administrative approval (and funding) were usually delayed and always apprehensive. . . . Even black college students who attend white colleges and universities have often established gospel choirs as an affirmation and continuation of their heritage."[7] Indeed, black gospel choirs are found at public and private colleges and universities all over the United States, including those whose student bodies are predominantly white. To designate the proliferation of collegiate gospel choirs as a movement seems appropriate because they emerged as student-initiated organizations during the peak period of black student involvement in public protests, political organizations, and demands for black studies programs and have outlasted many other institutionalized expressions of black awareness among college students.

The Howard University Gospel Choir was organized in 1969. At a worship service celebrating the choir's twenty-fifth anniversary in the spring of 1994, it was reported that two female students organized the choir after they each had dreamed of such a choir and shared their dreams with each other. At the time, students who majored in music were required to sing for the worship services at Rankin Chapel, in the university choir, whose repertoire excluded gospel music, clapping, holy dancing, or any other bodily movement with the music. Some students felt the need to sing music that reflected their own sense of religious and cultural authenticity, and the gospel choir gave them that opportunity. The choir quickly became a spiritual community that reflected some of the ecstatic worship and the moral rigorism of the Sanctified church tradition. Students were able to sing, clap, shout, and move with the music as they pleased. Several persons have testified that they were saved or miraculously rescued from danger while in the choir, after having experienced collegiate life as a period of seeking adult identity, rejecting their "church upbringing," being influenced by peers, and partying. The choir sponsored its own Bible studies, prayer meetings, and worship services. Clearly and unequivocally, the Pentecostal perspective emerged as a dominant factor in determining the direction of the choir and its repertoire.

Tension remains between Pentecostalism and more quietistic expressions of denominational Protestantism at Howard. Worship in Rankin Chapel has been supervised by three deans: Howard Thurman (1933–1944), Evans Crawford (1958–1993), and Bernard Richardson, who assumed the position in 1993. Thurman, a Baptist minister noted for his mystical approaches to religion, instituted interpretive dance and drama to enhance worship during his tenure.[8] Although his preferred worship demeanor as chapel dean was contemplative and quietistic, Crawford welcomed the participation of Pentecostal students in the religious life of the university. Bernard Richardson, an African Methodist Episcopal Zion clergyman, has pursued a similar policy of openness and inclusiveness with respect to Pentecostals at Howard. Gospel music was performed at his installation service on January 23, 1994, and several months later he invited the Howard Gospel Choir to provide the music for services at Rankin Chapel on one Sunday per month. This is significant because gospel music previously had been excluded from the chapel repertoire and from other official university-sponsored services and ceremonies as well. It took twenty-five years for the Howard Gospel Choir to be endorsed and accepted in this way.

The relationship between black identity and Pentecostal religion at Howard University has been addressed in an article by Pentecostal Chaplain Stephen N. Short, "Pentecostal Student Movement at Howard: 1946–1977." His records show that Pentecostal students first organized themselves at Howard in 1946 but did not receive official university recognition until 1966. Bishop Monroe Saunders of the United Church of Jesus Christ (Apostolic) was at the time a graduate student at the Howard School of Religion. He had organized the Pentecostal student movement at Morgan State University and Coppin State College. He persuaded a member of the divinity faculty to serve as faculty advisor to the United Pentecostal Association; the advisor subsequently refused to sign the necessary permission forms for the group to have a gospel concert on campus unless it could be guaranteed that "no one would shout, 'get happy,' or otherwise express himself in the prevailing emotional patterns common to the Pentecostal church."[9]

In the spring of 1969, Howard University was shut down by students as an act of protest. Thereafter, the student activity requirements were relaxed so that advisors did not necessarily have to be members of the Howard faculty or staff. The United Pentecostal Association then sponsored a highly successful gospel concert without the blessing of its faculty advisor. In response to the concert publicity, Short, a pastor of an independent storefront church in Washington, volunteered his services as faculty advisor. He organized the First Intercollegiate Pentecostal Conference at Howard in 1970, which drew 4,000 participants representing twenty-six states, including blacks, whites, Protestants, Catholics, Jews, classical Pentecostals, and charismatics. As a result of the conference, Dean Crawford formally accredited Short as Pentecostal chaplain of Howard University, the first such appointment ever made to a major college or university.[10]

In 1974, the name of the student organization was changed after an inquiry by the charismatic Episcopal priest Father Dennis Bennett, who wanted to know if the name "United Pentecostal" referred to the Oneness denomination. The students intended to convey the idea of Pentecostals being united and were unaware that the United Pentecostal Church was the name of a white Pentecostal body. To avoid confusion and to bring attention to the black founder of Pentecostalism, the students changed the name of their group to the William J. Seymour Pentecostal Fellowship of Howard University. In 1976, the fellowship purchased the first Pentecostal center in the world to be located at a major university, Seymour House, which "offers a full spectrum of services that include recreation, meals, counseling, Bible classes, seminars and workshops, emergency housing on a limited basis, and salvation through faith in Jesus Christ."[11]

It is evident, then, that both the gospel choir and the Pentecostal fellowship in its present form are related to the 1969 black student rebellion at Howard University. Students promoting racial justice and seeking free expression of cultural identity encountered the same forms of intransigence at black institutions as at white ones. That these developments should occur at a black university should come as no surprise, given the historical orientation of many black institutions of higher education to acculturation and assimilation into the dominant culture. As early as 1932, theologian Reinhold Niebuhr wrote about the tendency of Negro schools to encourage individual pursuit of self-realization without attacking the social injustices from which the Negro suffers. He was highly critical of the attitudes and behavior of educated Negroes with respect to the disinherited masses: "The progress of the Negro race, for instance, is retarded by the inclination of many able and educated Negroes to strive for identification and assimilation with the more privileged white race and to minimise their relation to a subject race as much as possible."[12] The push for gospel choirs, black student unions, black studies, and the like represented the black students' desire to identify with the "subject race" in the struggle for liberation and justice while enrolled in schools intended to prepare them for a more privileged existence. Black students at Howard demanded the institution of black studies programs in much the same manner as their peers at predominantly white universities, perhaps feeling their protest to be even more clearly justified on the basis of the existence of a black majority of students and faculty. Moreover, the general disdain many elite educated blacks have had for Pentecostalism remains problematic for Pentecostal students who have expressed feelings of rejection, ridicule, and unfair treatment on account of their religious beliefs and practices.

The marginalization of Pentecostalism on campus is certainly not unique to black higher education. One of the most important documents of the black student protest era, *Black Studies in the University* (1969), a book of proceedings of a conference on Afro-American studies at Yale, hardly mentions the black church or religion as a factor in black student consciousness or intellectual life. However, at Yale, a black worship service and a gospel choir were established during this time. At Yale and elsewhere, the gospel choir emerged as the core of black Christian identity for students who have felt marginalized on the basis of race, sex, culture, economic status, and religion. Noting that the "suspicion and disapprobation that have dogged Pentecostals throughout their history are still present in academic circles," theologian Harvey Cox commented that a Pentecostal student at Harvard was inspired to "come out of the closet" only after Cox's course on Pentecostalism appeared in the catalog.[13]

The proliferation of gospel choirs during and after 1969 included virtually all the historically black colleges and universities and most of the others. Collegiate gospel choirs quickly emerged as a national phenomenon. Some schools provided administrative support and rehearsal facilities for these choirs, as well as stipends for directors and accompanists. The Howard Gospel Choir recorded its first album, *Beginning,* in 1972, with endorsements from Delegate to the U.S. Congress Walter E. Fauntroy of the District of Columbia, Mayor Walter Washington of the District of Columbia, President James E. Cheek of Howard University, gospel artist Myrna Summers, and Howard alumna Pearl Williams-Jones. The Modern Black Mass Choir of Fisk University issued their *First Time Around* album on Nashboro's Creed label in 1973. This gospel choir was first formed in 1969 for the purpose of participation in a "week of reckoning" held on Fisk University's campus. In 1976, CBS Records released on the TSOP (The Sound of Philadelphia) label the *Third Annual National Black College Gospel Festival* as a two-volume album set featuring twenty-two collegiate gospel choirs.[14] One of the groups represented on the album, The Sounds of Blackness of Macalester College in St. Paul, Minnesota, has produced several highly successful recordings in recent years with significant cross-over appeal, buoyed by occasional appearances on nationally televised programs. At Morgan State University, gospel music was incorporated into the standard repertoire of the university choir. At Harvard University, the Kuumba Singers (Swahili for "creativity") present themselves as a black choir rather than a gospel choir, and their name avoids the direct Christian witness implicit in the term "gospel."

The University of Maryland Gospel Choir performed at the 1993 annual meeting of the American Academy of Religion, a scholarly gathering at which papers normally take priority over prayers and praise. In 1994, the Swarthmore College Alumni Gospel Choir performed for the annual alumni-alumnae collection, the first time the choir had ever been officially invited to appear before a college audience in its twenty-three years of existence. A member of the choir explained to the predominantly white audience that the choir and gospel music had special meaning for black students who felt alienated on campus because of their religion and their race. The songs performed were carefully selected to convey the underlying rationale for gospel music: praise of God and faith in Jesus Christ, with lyrics uncharacteristic of the standards of ecumenical and gender inclusiveness normally associated with the liberal arts college.[15]

The gospel choir movement in the United States reflects the general orientation of the Sanctified churches toward ecclesial reform, cultural resistance, and revolution or rebellion. The choirs have been ecumenical in composition and attracted students from a variety of religious backgrounds, even if the music and worship associated with these groups has been predominantly Pentecostal. Often students with no interest or involvement in religion join the choirs because of the music and the fellowship. Here ecclesial reform should be interpreted not in terms of the choirs' measurable impact on the established churches but rather as the representation of a viable alternative for students whose cultural and spiritual needs are not being met by established campus ministries, university choirs, or local churches. Despite the movement's alleged anti-intellectual biases and appeal to "ignorant" and uneducated blacks, the Sanctified church has traditionally encouraged its saints in higher education to acquire "the learning" without losing "the burning."[16] The gospel choirs have reinforced this exhortation in the minds of Holiness-Pentecostal students in pursuit of "the learning" while offering other students an initial exposure to black religious fervor, that is, "the burning."

The cultural resistance of black students to the pressure to assimilate the particular values, practices, and attitudes of the university setting and the societal classes it emulates has been a critical determinant of the success and distinctiveness of the gospel choir movement. These choirs provide black students with a space or context for worship and performance in relative isolation from the critique of individuals who do not comprehend or appreciate black culture. More important, the lyrics and rhythms of gospel music provide an aesthetically pleasing and spiritually enriched "kinetic vocabulary" (to borrow a term from Katrina Hazzard-Gordon), through which black students can articulate their collective needs, perceptions, impressions, and responses.[17] It is probably an exaggeration to designate the gospel choir movement as a revolution, but it represents one of the most vital ongoing institutional expressions of the ubiquitous student rebellions of the late 1960s.

In the heat of protest, it was not unusual for the more secularized and politically oriented black students to criticize the gospel choirs as counterrevolutionary. In retrospect, however, these groups have manifested black awareness and identity in concrete, visible forms. "I Believe," a selection from the *Africa to America* release by the Sounds of Blackness, was used as a campaign theme in Marion Barry's 1994 successful bid for a fourth term as mayor of Washington, D.C., after he had been convicted and imprisoned for using illegal drugs. The song echoes Barry's emphasis upon black redemption, identity, and empowerment. In the liner notes accompanying that recording is an explicit statement of exilic identity that is representative of the spirit of the collegiate gospel choir movement:

> We especially thank God for "Africa The Motherland," Her mighty kingdoms and royal people—our ancestors, whose faith, strength, perseverance and resilience empowered us to survive and create a family of musical styles that has profoundly influenced this entire planet. We are the true Kings and Queens of Swing, Rock & Roll, African melodies, Work Songs, Field Hollers, Spirituals, Jubilees, Blues, Ragtime, Jazz, Gospel, Rhythm & Blues, Soul, Funk, Reggae, Hip-Hop, Be-Bop and Pop. Our music is our life blood, our legacy, our heritage and our testimony. Therefore, we praise God for *all* of our music, that He alone has blessed us with to share with the rest of the world—on our terms.[18]

These choirs have provided a context for black students to give voice and rhythm to their exilic consciousness in the campus setting by singing the Lord's song in a strange land of intellectual, social, cultural, and religious alienation.

The Black Clergy Caucus Movement in the Church of God (Anderson, Indiana)

In the late 1960s and early 1970s, while black college students all over the United States were beginning to organize themselves into gospel choirs as a spontaneous act of collective affirmation of racial identity and religious commitment, a parallel development was taking place among black religious leaders. James Cone's *Black Theology and Black Power,* published in 1969, was the first attempt to articulate a black theology using the sources, style, and language normally associated with theology as an academic discipline. Its significance as a landmark theological statement has continually been acknowledged by black religious scholars since that time. It mirrors another category of collective response to racism informed by religious sensibilities and experience, the black clergy caucus.

The earliest and most important manifestation of the black caucus movement was the "Black Power Statement" published in the *New York Times* on July 31, 1966, by the National Committee of Negro Churchmen (later known as the National Conference of Black Churchmen). In his introduction to this and other related documents in *Black Theology: A Documentary History, 1966–1979,* Gayraud Wilmore acknowledged that this document was "the turning point in the history of Black Church involvement in the civil rights movement"; from it was derived a theological analysis that sets the stage for the emergence of black theology.[19] Wilmore denied that the primary emphases of black theology were extrapolated from it, but Cone both quoted and discussed the "Black Power Statement" in *Black Theology and Black Power.*[20] Wilmore related the formation of black caucuses in several denominations during the late 1960s with the "first glimmerings of black theology." Black caucuses were organized in at least ten of the largest predominantly white Protestant church bodies and in the Roman Catholic Church; "roughly from 1965 to 1970, Black clergy and laity who had been thought to be securely 'integrated' in the White Church reconstituted themselves in separate Black enclaves and declared war on the White establishment and its theology."[21]

Black evangelical, Holiness, and Pentecostal clergy also participated in the caucus movement. Two documents included in the first volume of *Black Theology* are essays written by black evangelicals, whom Wilmore saw as "representative of a new mood among the younger clergy of Black Pentecostal and Holiness churches."[22] Efforts by leaders of the National Black Evangelical Association (including politically active Pentecostal clergy such as Herbert Daughtry and William Bentley) to create a united front with the National Conference of Black Churchmen ultimately failed.[23] However, there were Holiness-Pentecostal black clergy who participated in the black caucus movement and who appropriated the black power concept in terms consistent with the experiences, perspectives, and traditions of the Sanctified churches. The example presented here for closer scrutiny is the black clergy caucus of the Church of God (Anderson, Indiana).

On April 13 and 14, 1970, the Caucus of Black Churchmen in the Church of God was convened in Cleveland, with ninety-two delegates. (Although the term "churchmen" was used and the authors of the individual papers and statements are all male, clergywomen were present and generally have been included in all official gatherings of black clergy in the Church of God.) The proceedings were published in *The Church of God in Black Perspective.* The purpose of this caucus is clearly stated in Ronald J. Fowler's introduction, which is indicative of the church's characteristic orientation toward spiritual discernment and sense of divine call: "It has been convened to allow the victims of racial oppression affiliated with the Church of God to determine for themselves their assignment under God amid the black revolution."[24]

The ad hoc steering committee set forth a fourfold statement of purpose emphasizing the projected impact of the caucus proceedings in the Foreword: (1) to lead the church into a "more authentic and vibrant life and witness," (2) to serve as light to lead God's people forward across paths of "brotherhood and righteousness;" (3) to give impetus to those efforts and ministries designed to serve the total "man," and (4) to rally the church "to give birth, however painful, to a united church for a divided world."[25] The expression "a united church for a divided world" was taken from the church's international radio broadcast, "The Christian Brotherhood Hour," and cited deliberately by the black clergy to highlight the failure of the broadcast and the larger church body to give adequate attention to racial concerns.

Several papers were presented by leading black clergymen in the Church of God in the published proceedings. These pastor-scholars, most of whom are seminary graduates, cited theological and historical works by black authors, including Cone's *Black Theology and Black Power,* Lerone Bennett's *Before the Mayflower,* Thurman's *Jesus and the Disinherited,* and Frazier's *The Negro Church in America.*

The major findings of the meeting were issued in the form of several statements. The initial statement of position began with a strong affirmation of the National Association of the Church of God, the fellowship of black churches and ministries that convenes annually at West Middlesex, Pennsylvania:

> We proclaim to all that the National Association MUST and WILL remain— as a center of fellowship for people; especially the disinherited; — as a creative outlet for ministries especially relevant to the larger black community and thereby broadening the ministry to the total church; — as a training base for the development of black leaders; — as a school, forum, and social laboratory for black churches, and — as a place where the Gospel is shared.[26]

The statement of progress attempts to acknowledge the efforts by the Church of God agencies at Anderson, Indiana, to include black representation and to address black concerns. Suggestions are made for improving race relations, with a view toward helping the church "to achieve a balance between preachments and practice." Last, three statements of concern address specific issues: church investments, support of the National Association, and lay training programs in evangelism. There is a call for a moral audit of the church's financial investments, based on the belief that "none of the Church's funds should be invested with institutions, firms, organizations, corporations, or other interests whose practices and policies are racist or supportive of racist ideologies."[27] Next is an appeal for support of the National Association through the world missions budget of the Church of God and by individual black and white congregations. The third concern is a critique of the racially biased evangelistic training provided during the 1970 Church of God convocation by the white conservative evangelical organization, Campus Crusade for Christ, International.

The final document is the text of a sermon, "The Face of Jesus," preached by James Earl Massey, then campus pastor of Anderson College, who later became dean of the Anderson School of Theology. He is one of the leading homileticians and New Testament scholars in the Church of God and the author of several important books, including *Designing the Sermon* (1980), *The Hidden Disciplines* (1972), *The Responsible Pulpit* (1974), and *The Sermon in Perspective* (1976).[28] Citing 2 Corinthians 4:6 as a Scripture text, Massey described the face of Jesus Christ with reference to the ethnicity, suffering, and determination of Jesus as Savior. Massey's sermon bears the influence of his mentor, Howard Thurman, whose christological perspective presented in *Jesus and the Disinherited* (1949) predates the emergent black christologies of the black theology movement by some twenty years. Massey drew upon current events and personalities associated with the black consciousness movement—the 1968 Detroit race riot, James Baldwin, Albert Cleague, and Mahatma Gandhi—to illustrate the point that human lives can either reflect or obscure the splendor in the face of Jesus. One particular incident he recalled from the Detroit riots effectively illuminates the question of the blackness of Jesus Christ on several levels:

> My home city of Detroit was shocked two years ago when a gray stone statue of Jesus, standing on the lawn of Sacred Heart Seminary—just across the street from my church—turned up black during the Detroit riot. . . . The statue had been painted black by two angry black boys from the community who inwardly sensed that the real Christ is for all, that he belong[ed] to their black community, that he gathers up their particularity in his own life, that he accepts them, too. It was an act of self-affirmation. It was also an act of insight into Jesus as the one with the special face—the face that Western life and culture has obscured. The officials at Sacred Heart caught the point and let the statue remain black. They began to sense their own hang-up in chromatics and culture, having forgotten that Jesus was identified initially with the poor, the needy, the harassed, the disinherited. Those officials rightfully took another look at the true face of Jesus. . . . That black paint was prelude to a fresh hearing of the gospel.[29]

Massey interpreted this act of vandalism as black youth making a christological statement of self-affirmation. For the church officials, the incident forced the issue of reconsideration and revision of the white Christ created by Western culture. To the hearers, it represented an opportunity for a new rhetoric to be heard with new ears. For this group of Holiness clergy, preaching had always been the ultimate forum for adjudication of theological ideas and political concerns. Thus, this final homiletical treatment of the central themes of the black caucus represents the highest level of expression and assent that can be ascribed to any issue.

One of the papers in the main body of the proceedings is also a homiletical statement, a sermonic defense of the National Association of the Church of God that exhorts black clergy and congregations to unity in the face of white Christian racism. Benjamin F. Reid, then pastor of the Southwestern Church of God in Detroit, used a critical, confrontational style of address in his contribution, "The National Association Must Live On!" Reflecting on the early history of the National Association (as described more fully here in chapter 2), he asserts that it is not a separate communion but has a distinct mission and purpose:

> Let us remind ourselves that we are not the Black Headquarters of the Church of God. We are not a competitive organization with Anderson, Indiana. We are not a black activist political movement nor are we a front for any anti-American, hate centered, violence prone black anarchist outfit. We are not the foundation for a black Church of God denomination or movement. We are simply an organization that grew out of the need for fellowship among the black saints without racist overtones and segregationist control. We are here because there needed to be a place where black leadership could be developed and exercised. We are here because there needed to be an outlet for black religious creativity, spiritual development and emotional freedom. We are here because in 1917 as well as in 1970 America feels that white is right, white patterns are the only acceptable patterns of behavior, and white methods are preferable to any other kind. We are here because, sad to say, in the Church of God, as in the rest of American society, Black Unity is the only way to survive in the midst of a racist culture. [30]

He saw the National Association as a fellowship of black clergy and congregations that has been a rallying point for protest, an educational and experimental institution for black leadership development, a religious-political power base, and "a place where we worshipped God and heard in our language and according to our cultural patterns the great message of the Church of God."[31] In addition to these factors that highlight the importance of the National Association to blacks, its impact on the Church of God as a whole was bluntly stated: "Without the Association, the Church of God, Anderson, Indiana would be just like the Pentecostal Holiness Church, the Church of the Nazarene and the Assembly of God—basically lily-white, middle class, holiness church groups, reflecting the sin and shame of Racist America!"[32]

Reid's description of the identity and contributions of the National Association reflected an exilic perspective that holds that blacks must unite in order to survive and to maintain a critical, ethical dialogue with white Christians. He lifted up this black organization as a progressive model of black unity, identity, and success that could help to meet the needs of black youth in an era of black consciousness:

> At a time when black youth are looking for black-oriented, black-run and black-understanding institutions, this can be our finest hour. While we do not, cannot and will not preach separation, hatred, violence and militantism, yet we can stand as a living evidence that right living, God-fearing black men and women can achieve and succeed and under God make society a better place.[33]

With regard to relations with white Christians, Reid complained that because white Church of God clergy have become theologically liberal, doctrinally uncertain, and biblically "sterile," they needed the "strong biblical emphasis" cultivated by the black clergy of the National Association. He commended the association to whites as a model of integration, whose personnel, participation, and programming have always been inter-racial. The dialectical relationship between black and Christian identity is also the watershed of authentic Christian witness:

> We then give witness to the world that Christian behavior and Christ-like ethic pervades and supersedes our devotion to blackness. The National Association can be an expression of black identity and black unity, and it can also be an expression of Christian witness that breaks down walls of injustice and preju-dice in order to bring white and black Christian into *true* fellowship.[34]

Reid saw the Church of God needing a "theological injection" from its black churches, who hold the key to spiritual renewal. He suggests that a renewed emphasis on the practices of the black churches—"getting folk saved," "getting people filled with the Holy Spirit," "practicing divine healing," "demanding that the saints be different from the world," and "letting the Lord have his way in the service"—throughout the entire Church of God can bring about a "real revival."[35] The features listed here are, to be sure, the liturgical and ethical distinctives the black Saints in the Church of God hold in common with the Sanctified church tradition in general.

As a whole, the proceedings of the Caucus of Black Churchmen in the Church of God reveal dialectics of protest and cooperation similar to that described by Gilkes. This dialectics is driven by two factors: (1) the black struggle against discriminatory structures and patterns of subordination based on race and (2) the determination to maintain unity with white Christians in the face of racism and discrimination. In 1970, the black Church of God clergy were engaged in reform, resistance, and rebellion. From the beginning, the Church of God has understood itself to be a "reformation movement" opposed to the divisiveness of denominationalism. The term "Church of God Reformation Movement" is still used today in some circles as the preferred designation for this religious body as a whole. The pursuit of reform in the area of race relations became an important facet of the broader task of bringing the Christian church into closer conformity with the demands of the gospel. The posture of resistance and rebellion is forcefully articulated in the paper by Charles Pleasant: "Racism and all its ugly manifestations are continually warring against Christ and the black church. The stand of the black Church is rebellion."[36]

In sum, the commitment of these black clergy to the fundamental tenets of the Church of God—that is, to the doctrines of holiness and unity—mandates the pursuit of racial reconciliation through confrontation and protest. The responses of the Church of God black clergy caucus to the black power challenge were grounded in a long-standing tradition of black self-determination nurtured by an organization that was created decades before as a platform for worship, spiritual formation, and leadership development, not as a mere reaction to racism or as a parallel, independent structure like the black Baptist and Methodist denominations. The concerns of the black clergy caucus of 1970 had been addressed as early as 1917, when the West Middlesex camp meeting was established to foster and preserve Sanctified worship practices, preaching, piety, and social ethics. In this light, the Church of God black clergy caucus provides an interesting and illuminating example of the balance of communal and unilateral power that Peter Paris describes in *The Social Teaching of the Black Churches* with reference to the black Baptists and Methodists.[37]

The push toward the increased sharing of power and resources with whites in the Church of God without compromising black identity and self-interest has paid off in many ways. Of the seven authors listed in the table of contents of the proceedings, four have held executive positions with considerable power and visibility in the Church of God: Sethard A. Beverly as director of Metro-Urban Ministries; Edward L. Foggs as general secretary of the Church of God, International; James Earl Massey as dean of the Anderson School of Theology and, for several years, preacher for the church's syndicated weekly radio broadcast, the "Christian Brotherhood Hour"; and the late Thomas J. Sawyer as staff associate in missions. The remaining three authors have held important urban pastorates—Robert O.

Dulin in Detroit, Marcus H. Morgan in Chicago, and Benjamin F. Reid in Los Angeles—and served the church at large in appointed and elected posts. The black clergy have held midyear meetings annually since 1970 to give vigilant attention to a host of issues and dilemmas relevant to the black churches and their constituents.

The current mission statement of the National Association of the Church of God emphasizes Christian ministry and identity in nonracial terms and describes the organization as

> A voluntary association of Christians and churches affiliated with the Church of God (Anderson, IN) who, guided by the vision of Elisha Wimbish, mobilizes ministries and resources through national forums to remind, inspire, renew, equip and enable persons to live in obedience to Jesus Christ, primarily with an annual camp meeting at West Middlesex and other auxiliary functions throughout the year.[38]

Thus, the orientation toward reform, resistance, and rebellion in relation to whites is not the bedrock of identity for these black Christians but, instead, is the outgrowth of an eschatological community guided by a vision of racial harmony, spirited worship, and holy living.

6

Black Intellectuals and Storefront Religion in the Age of Black Consciousness

Black Intellectuals and Black Religion in Exilic Perspective: A Typology of Traditions

By definition, exile is the experience of being a stranger or outsider far removed from home. In the conceptual language and context of the black intellectual, exile is a way of describing the predicament and state of mind of critical thinkers who self-consciously maintain a sense of racial and cultural identity with another country or nationality. For the African American, it implies a peculiar "both-and" dialectic, identifying the self as *both* American *and* African, as a permanent resident of an alien land.[1]

In his book *Prophesy Deliverance!* philosopher Cornel West presented a typology of black intellectual traditions that rests on diverse depictions of Afro-American religion and takes a significant step toward formulation of a "unified theory" of black culture and religion.[2] West outlined four traditions, drawing his examples from the ranks of notable twentieth-century black intellectuals and writers: the exceptionalists (e.g., W. E. B. DuBois) laud the uniqueness of Afro-American culture; the assimilationists (e.g., E. Franklin Frazier) consider Afro-American culture and personality to be pathological; the marginalists (e.g., James Baldwin) see the culture as restrictive, constraining, and confining; and the humanists (e.g., Ralph Ellison) extol the distinctiveness of the culture while affirming that Afro-Americans are not above or below the rest of the human race.[3] Each of West's categories rests on a particular depiction or evaluation of black religion. This typology is used as a guide to interpreting the thought of a select group of black intellectuals of the black consciousness era who in one way or another addressed the problem of exilic existence in light of the symbols and meaning of the black folk religion as practiced in slavery and in the urban storefronts of the twentieth century.

The exceptionalist perspective is presented in one of the most important documents of the black consciousness era, Chancellor Williams's *The Destruction of Black Civilization: Great Issues of a Race from 4500 B.C. to 2000 A.D..*, published in 1976. This book was an ambitious attempt by a Howard University history professor to document the history of the African people based on sixteen years of international research and field studies in a single volume, "so written that Black John Doe, cab driver or laborer, and Jane Doe, house maid or waitress, can read and understand the message from their forefathers and foremothers as well as college students and professors."[4] His purpose was to explain the oppressed condition of modern blacks by describing how the ancient African civilization was systematically destroyed by whites.

Williams's inquiry started from a personal childhood remembrance of poverty and racial segregation in the rural South:

> In a small town surrounded by cotton fields in South Carolina, a little black boy in the 5th grade began to harass teachers, preachers, parents and grandparents with questions none seemed able to answer: How is it that white folks have *everything* and we have *nothing?* Slavery? How and why did we become their slaves in the first place?[5]

He surveyed 6,000 years of African history from the rise of Egyptian dynasties to the fall of the last indigenous empire in the southern part of Africa in 1902, briefly addressed the problematic relations between the races in the United States, and then concluded his study with a detailed "organization for unity" plan designed to foster black progress.

Williams's concern for making his scholarship accessible to working-class blacks, although somewhat unusual for an academician, is less surprising in view of his own background and his familiarity with the plight and aspirations of the black poor in the urban storefront churches. His 1949 doctoral dissertation was a study of the socioeconomic significance of the storefront church movement. There he addressed the problem of class discrimination among blacks in the major Christian denominations as a factor giving rise to storefront churches that are more willing to welcome and minister to the disinherited:

> And, as it has been shown, the problem is not racial. Very little is ever said about the marked and widespread discrimination and class segregation *within* the Negro race. Some Negroes feel as much out of place in certain Negro church groups as they would in any of the major white church groups. The "ordinary" colored person is not wanted in the upper class Negro church.[6]

Although the dissertation and the book differ greatly in scope, content, and style, the book's conclusion uncannily echoes the perspective of the dissertation with regard to the problem of race and class discrimination among blacks in the United States: "Now it is just here *within the race* where *integration* is not only needed, but it is mandatory. We shall remain a weak people until we begin the drive for integration of Blacks first of all, instead of battling to integrate with other peoples."[7]

Williams's insightful application of the concept of integration to intragroup relations exposes the moral dilemma faced by blacks who press for acceptance by whites while at the same time distancing themselves from other blacks. The fact that he would raise the same question in 1976 that he first posed in 1949 indicates that the black consciousness movement had had limited impact on the persistent patterns of discrimination operating in black institutions, including the Christian churches. Williams was not attacking Christianity as an outsider or skeptic. He described himself as a devout Christian honestly facing up to the moral failures of other Christians: "This writer, for example, is a devout Christian, but that fact does not blind him to the chains of bondage hammered on his race in the name of Christianity or cause him to try to gloss over or soften the records of history when his own religion is unfavorably exposed."[8] The future task he envisioned for black Christians (and Muslims) was to distinguish "true" religion from its evil uses by applying critical ethics to religious faith and practice as an impetus to black unity. Williams challenged black Americans, as a people in exile, to participate in the black revolution on the basis of an honest appraisal of their own complicity with the historical role played by whites in the destruction of black civilization and the distortion of Christian ethics.

A second representation of the exceptionalist tradition was developed by anthropologist Leonard Barrett, author of *Soul-Force* (1974). He observed a trend toward the revival of black pride, a revival echoed in slogans such as "Black is beautiful," "Black Power," and "Black theology." There was a return to distinctive Africanisms such as the natural hairdo, African robes, African names, and the increasing frequency of black pilgrimages to the motherland. Barrett predicted that historians would come to see this "Age of Black Awareness" as an era when "Africans in 'exile' returned to themselves," accomplishing a black cultural emancipation sought by struggling blacks for more than 300 years.[9]

Barrett's *Soul-Force* was intended as an interpretation of the concept of "soul" (or what he calls "soul-force"), which had became an extremely popular expression for the uniqueness of black people during the 1970s. He set out to prove that the African brought a rich cultural heritage into the

New World and, in the new environment, remained a variation or modification of the African self.[10] Thus, the cultural emancipation of the African in exile is understood to be a spiritual quest.

The assimilationist perspective of sociologist E. Franklin Frazier is evidenced in his influential book *The Negro Church in America,* published posthumously in 1964. Frazier represents an interesting example of the dialectics of black intellectual exile. He was genuinely committed to integration and assimilation as appropriate goals of black progress, yet he focused much of his scholarly attention on the study of black institutions and spent most of his teaching career as a professor at the predominantly black Howard University. The content of the book is based on an interpretation of black religious and social life as originally offered to a European audience; it is the revised text of his 1953 Frazer Lecture in Social Anthropology given at the University of Liverpool. Many of his black contemporaries sought European audiences for their intellectual and artistic reflections on black life.

The book begins with a now famous challenge to anthropologist Melville Herskovits's view that black American culture is shaped by the African cultural background, as set forth in his 1924 study, *The Myth of the Negro Past:* "In studying any phase of the character and the development of the social and cultural life of the Negro in the United States, one must recognize from the beginning that because of the manner in which the Negroes were captured in Africa and enslaved, they were practically stripped of their social heritage."[11] This debate is rooted in the two distinctive schools of thought with which the two thinkers had become identified. At the University of Chicago, Frazier had been a student of Robert E. Park, who originated the theory of social disorganization, whereas Herskovits had studied at Columbia University with Franz Boas, a proponent of cultural relativism. Herskovits viewed the behavior of black Americans as culturally *different,* a product of their distinctive African background. Frazier saw black communal life as culturally *deviant,* the result of imperfect assimilation of American social norms because of poverty, ignorance, oppression, and the heritage of slavery.[12] Although subsequent studies of black religion and culture have made much of the debate between the two, Frazier settled the debate himself no sooner than setting it forth in a note that quotes Herskovits as saying that New World African cultural survivals were the least intense and the most generalized in the United States.[13] In other words, their disagreement was not a fundamental dispute as to whether Africanisms carried over into black American life; it is more a matter of assessing the specific tribal and regional character of African cultural practices and the extent of their influence in the United States.

Frazier appeared to be somewhat inconsistent in his social analysis of the Negro church against a backdrop of African influences. On the ground of his theory of Negro social and cultural deviance, he claimed that it is "impossible to establish any continuity between African religious practices and the Negro church in the United States." Yet, he admitted that the "shout songs" and "holy dance" performed in the storefront churches "most likely reveal a connection with the African background."[14] Another point of contradiction or ambivalence can be discerned in his discussion of black sacred folk songs. Frazier took the position that the spirituals sung by slaves were essentially "religious" in sentiment and otherworldly in outlook, and he insisted that Negro intellectuals have erred in attributing revolutionary meaning to them. However, he did not adhere to his own advice with regard to the analysis of black religious practices in general and so ascribed a social function to every aspect of black church life. He condemned black scholars for allowing themselves to be influenced by white radicals who encourage political interpretation of spirituals, but the white influences on his own sociological thinking evoked no similar critique. If the spirituals are more properly understood to be essentially "religious," then why not also refrain from the functional interpretation of other "religious" practices? Sociologist Harold Dean Trulear has noted the pitfalls inherent in this functionalist approach to interpretation of the ideational structures of black religion (i.e., belief, symbol, and theology) principally as products of the social structures of slavery, segregation, and urbanization; he proposed a more holistic sociological perspective that enables the interpreter of black religious ideation to take into account factors other than black responses to the white social context.[15]

On the surface, Frazier's description of the Negro church community as "a nation within a nation" for the socially and morally isolated masses of black Americans as a refuge in a hostile white world suggests the notion of exile. Because his analysis was so severely skewed toward the ideals of integration and cultural assimilation, however, this "nation within a nation" was understood strictly in terms of the victimizing effects of segregation. He condemned the Negro church as "the most important institutional barrier to integration and the assimilation of Negroes" in an ironic disregard of the roles the churches were playing in the rise of the civil rights movement in the South.[16] Frazier concluded his study with an indictment of the Negro church because of its intimidation of intellectuals and its responsibility for the "backwardness" of American Negroes. The key to black intellectual development was seen as integration into American institutions and "escape" from the "stifling domination" of the Negro church; Frazier boldly predicted the demise of "the social organization of the Negro community, in which the church is the dominant element." Progress would be achieved to the extent that these structures crumble "as the 'walls of segregation come tumbling down.' "[17]

Exilic existence became a sad and tragic scenario in Frazier's perspective: "Negroes are increasingly cast afloat in the mainstream of American life, where they are still outsiders," with no hope of acceptance by whites.[18] However, the Negro church was no solution to this dilemma, insofar as its importance was understood solely in terms of the restricted participation of Negroes in American society. In other words, the churches are no longer needed by those who have acquired social status, but the churches and social organizations of a racist society reject even the black middle class. Those persons desiring to become assimilated and acculturated encounter in the dominant society cultural values and practices that foster denigration of the black self seeking acceptance. Yet, accommodation and social isolation cannot present a satisfactory alternative for the exile who sees black institutions as inferior, dysfunctional, and deviant. Thus, the black intellectuals who despise black institutions and culture but are themselves despised by the dominant culture have no place to call home. Moreover, if exile is understood and depicted in an atheistic frame of reference devoid of God and religious faith, it can no longer be regarded as a refuge for the soul but is experienced as something much worse. As Frazier once wrote: "Man is the only divinity we know and need to know. . . . He has eaten of the Tree of Knowledge and some day may eat of the Tree of Life. Earth will become his only heaven and exile his only hell."[19]

The marginalist tradition would seem to be the most critical of the four as a category for illuminating the concept of intellectual exile. It is represented well by James Baldwin, one of the leading writers of the black consciousness era. Although Baldwin's most important essays on black life were published in the 1960s, he remained a significant force in African American letters until his death in 1987. His special relevance to this discussion is his deep rootedness in and rejection of the Sanctified church. Moreover, he exemplifies an extreme posture in black intellectual exilic consciousness because he eventually left America to live in self-imposed exile in Europe.

Baldwin was literally a child of the Sanctified church; his father was a storefront preacher. However, the young Baldwin despised his father, whom he experienced as a strict and abusive parent struggling to raise his children in the abject poverty and crowded conditions of crime-ridden Harlem. West heard the rhythm, syncopation, and appeal of black preaching in Baldwin's moralistic essays on the problem of race in America. Baldwin had spent a brief period during his adolescent years as a preacher, and the salient values of his subsequent homiletical discourse are the familiar Christian ones: love, mercy, grace, and inner freedom. His "candid acceptance of personal marginality to both Afro-American culture and American society" and his "moral sermonizing to all Americans" are expressive of a distinct mode of exilic consciousness that nevertheless echoes both Frazier's contempt for black folk religion and his desire for integration and acceptance.[20] Like Frazier, Baldwin "overlooks the possibility of cultural vitality and poverty-ridden conditions existing simultaneously in Afro-American life."[21] Instead, the most negative and culturally stifling aspects of black culture seemed for him to be embodied in the materially impoverished urban storefronts of the Sanctified church tradition.

Thus, Baldwin exiled himself to the margins of a black culture whose richness, beauty, and eschatological hope had formed and nurtured him:

> Baldwin did not abhor this culture; he simply could not overlook the stifling effects it had on nonconformists. He wanted desperately to identify with Afro-American culture, but he took seriously the Christian, humanist values it espoused and the artistic imagination (the nonverbal or literate expressions) it suppressed.[22]

Despite his efforts to distance himself from the Sanctified church, Baldwin describes with great eloquence its permanent stamp and imprint upon his being:

> The church was very exciting. It took a long time for me to disengage myself from this excitement, and on the blindest, most visceral level, I never really have, and never will. There is no music like that music, no drama like the drama of the saints rejoicing, the sinners moaning, the tambourines racing, and all those voices coming together and crying holy unto the Lord.[23]

Baldwin's biographer and secretary, David Leeming, revealed that "Crying Holy" was the original working title of Baldwin's best-selling autobiographical novel, *Go Tell It on the Mountain*. Although this earlier label is one of the characteristic expressions of the Sanctified worship experience, the title finally given to the book was taken, somewhat arbitrarily, from one of the Negro spirituals. Leeming cited Eleanor Traylor of Howard University to establish "the Baldwin narrator-witness" as a feature of Baldwin's fictional writing. Baldwin wrote that he had "come through something" the morning he finished typing the manuscript, suggesting that once he had given voice to his Sanctified conversion experience, he underwent a further conversion or call to the literary task of "witnessing."[24]

Leeming also lifted up the religious notion of confession as an essential theme in Baldwin's life and work:

The refusal to risk the danger of confession was at the base of the great social failure revealed by American racism. Only by letting go, by entrusting one's life to another, could an individual or a society hope to achieve wholeness. The doctrine of confession was a staple at the real-life Baldwin welcome table, as it is in the play.[25]

The play is *The Welcome Table,* a slice-of-life work about exiles and alienation that was completed in rough form the year Baldwin died. Reflecting on Baldwin's death in the company of friends in an apartment at Saint-Paul-de-Vence in France, Leeming drew upon a host of religious metaphors to characterize the relationship between the play and Baldwin's exilic experience:

It was rooted metaphorically in the welcome table of the other world, where the weary traveler would find "milk and honey" and in the welcome table/ altar of the church that had once been Baldwin's, where the sacrificial victim becomes food for the "children," for the "saints" of the "amen corner" and the "hallelujah chorus." It could serve now as a metaphor for the work in which his "gospel" or prophecy was celebrated and around which the "brothers" and "sisters" of his communion congregated, as they actually had done in the house at Saint-Paul. The welcome table was a place of witness, where exiles could come and lay down their souls.[26]

As a final affirmation of the abiding meaning of the Sanctified church tradition in Baldwin's life and artistry, Leeming reported that a tape recording of Baldwin singing "Take My Hand, Precious Lord" was played at his funeral in New York: "It startled the listeners. He seemed to be there, still witnessing, and people were moved."[27]

The fourth category of black intellectual response that emerged during the 1970s as an alternative to the exceptionalism of Williams and Barrett, the assimilation espoused by Frazier, and the marginalization of Baldwin in exile was the humanist tradition. In West's view, this tradition is best exemplified by African-American music:

The rich pathos of sorrow and joy which are simultaneously present in spirituals, the exuberant exhortations and divine praises of the gospels, the soaring lament and lyrical tragicomedy of the blues, and the improvisational character of jazz affirm Afro-American humanity. These distinct art forms, which stem from the deeply entrenched oral and musical traditions of African culture and evolve out of the Afro-American experience, express what it is like to be human under black skin in America.[28]

This assessment points to the work of two black theologians who gave attention to the analysis of black religion and black music in a humanistic frame of reference: James Cone and Howard Thurman.

Cone's 1972 *The Spirituals and the Blues* attempted a theological analysis of materials drawn from black folk culture, both religious and secular. He mined the universal themes and theological claims inherent in this music. The ontological statements about black being and black life are characteristic of his other theological writings, but he also discovered in the spirituals and the blues affirmations of truth, hope, and freedom with a universal appeal.

Cone traced his own experience of "the power of song in the struggle for black survival" to the place of his own upbringing, Bearden, Arkansas, and the Macedonia African Methodist Episcopal Church. Although the claim that "black music must be *lived* before it can be understood" is offered retrospectively by Cone with reference to his past, as a black intellectual he generally afforded a more positive endorsement and legitimation of black folk culture than either Frazier or Baldwin.[29] However, he had in common with Baldwin a wealth of past personal memories of participation in the black church from which to draw insights into black folk culture, and he shared Frazier's penchant for functionalist interpretations of black religion.

This point is more clearly demonstrated in his 1981 essay on black worship, in which Cone's functionalist approach is applied to two aspects of black worship peculiar to the Sanctified church tradition: the ritual dance known as shouting and the Wesleyan doctrine of sanctification, or holiness. He insisted that the ritual of shouting be interpreted strictly in terms of racial identity and social status: "The authentic dimension of black people's shouting is found in the joy they experience when God's Spirit visits their worship and stamps a new identity upon their persons, in contrast to their oppressed status in white society. This and this alone is what the shouting is about."[30] Next, his definition of the experience of sanctification is rendered as exclusively expressive of the struggle for liberation and the pursuit of a radical political vision:

> Sanctification in black religion cannot be correctly understood apart from black people's struggle for historical liberation. Liberation is not simply a consequence of the experience of sanctification—sanctification *is* liberation—that is, to be politically engaged in the historical struggle for freedom. When sanctification is defined in that manner, it is possible to connect it with socialism and Marxism—the reconstruction of society on the basis of freedom and justice for all.[31]

It appears that Cone has completely misread the self-understanding and worship rituals of the Sanctified church tradition. His Marxist understanding of sanctification reflects once again the tendency of black intellectuals to import theories and perspectives assimilated from Europe or from white

America to interpret materials from their culture and/or experience. Cone relentlessly employed a dialectic of oppression that reduces all religious, cultural, social, and political activities of black people to acts of resistance directed toward their white oppressors. He sorted out black cultural forms for examination and interpretation on the basis of whether they present grist for the liberationist theological mill.

It is not surprising that Cone and other black liberation theologians have struggled to find acceptance in the black churches. In his recent work on African American spirituality, Theophus Smith concluded that liberation theology tends to displace black people's multifaceted religious experience with a one-dimensional interest in a hermeneutic of liberation: "Black religious experience, however, is not reducible to the experience of suffering and oppression, nor to the quest to overcome suffering and oppression."[32] He noted that black theologians have yet to hold white academic approaches in creative tension with ancestral and indigenous modes of black thought.[33] Their intellectual marginalization vis-à-vis the black churches is directly attributable to their selective and skewed approaches to black religious and cultural data and to their tendency to discard, ignore, or misinterpret data that manifest meanings other than the black struggle against white oppression. As Lincoln and Mamiya observed in their 1990 sociological study of the black church, black liberation theology has had limited influence on black urban ministers in the United States; their data suggest that two-thirds of these ministers have not been affected by it at all.[34]

During the 1970s, however, Cone's work attracted a great deal of attention because of his willingness to engage in theological terms the slogans and ideology of black power at the height of its flowering. His contribution is important because the emergent black power movement was largely secular in character and, in many respects, entailed a strident rejection of religion and religious approaches to the struggle against racist oppression. In his last book, *Where Do We Go from Here: Community or Chaos?* (1967), Martin Luther King Jr. assessed the spiritual strengths and liabilities of the black power struggle; Cone's 1969 *Black Theology and Black Power* furthered the dialogue between the black church and black power. He continues to be a dominant force among black religious scholars and is greatly to be admired for his openness to growth and critical response.

As an humanistic interpreter of black religion and culture, Howard Thurman represents an interesting and illuminating example of the black intellectual experience of exile for several reasons. He experimented with ways of unifying theory and practice (1) by becoming cofounder and copastor of the Church for the Fellowship of All Peoples in San Francisco from 1944 to 1953, a church heralded as the first authentically inclusive model of institutional religion in the United States, and (2) by his ministry to students as a chapel dean who employed innovative approaches to worship involving dance, drama, and the visual arts at Howard University and at Boston University. Thurman remained vitally engaged in worship and church life, not as a distant or despised childhood memory but as an ongoing priority and calling. In view of the fact that Thurman remains the single most prolific writer on the subject of religion among African American scholars—some twenty-two books and forty-five articles—one is all the more impressed by his commitment to preaching as a vehicle for his ideas. He seems to have made a unique contribution as a religious thinker by his insistence on communicating his ecclesiological vision to general audiences, based on his extraordinary ability to wield both the spoken word and the written word with equal power and grace.

In 1975, Thurman reissued *Deep River and the Negro Spiritual Speaks of Life and Death,* a combination of two lecture series that originally had been published in the 1940s. It is a poignant interpretation of the religious meaning of the music and the self-understanding of the slaves. Thurman's somewhat unorthodox academic writing is more poetic than pedantic, and the reader becomes painfully aware of his apparent aversion to documentation of sources and references (especially of the folk materials). Moreover, on occasion, he adopted a defensive tone in response to the critique of some youthful activists that he was too mystical and quietistic. Although he was chided for not participating in the vanguard of the civil rights movement and the black power rebellion, he was a part of the historic 1963 March on Washington. In a general introduction to the 1975 edition, Thurman noted the relevance of his studies of the spirituals to the times and the consistent demand for them to remain in print:

> This demand was greatly intensified during the period marked by a fresh sense of root or collective self-awareness brought into sharp focus by the tempests of the Civil Rights Movement. Despite the primary secular and political character of the movement it found sources of inspiration and courage in the spiritual insights that had provided a windbreak for our forefathers against the brutalities of slavery and the establishing of a ground of hope undimmed by the contradictions which held them in tight embrace.[35]

Thurman brought great skill and sensitivity to the task of interpreting the multiple messages of the spirituals, not in terms of a functional analysis of their legacy of protest and resistance but, rather, with attention to the full humanistic impact of lyrics and rhythm and performance. He invited the reader to reflect on the Negro spiritual "We Are Climbing Jacob's Ladder":

Have you ever heard a group singing this song? The listener is caught up in the contagion of a vast rhythmic pulse beat, without quite knowing how the measured rhythm communicates a sense of active belonging to the whole human race; and at once the individual becomes a part of a moving host of mankind. This is the great pilgrim spiritual.[36]

In a brief foreword, "On Viewing the Coast of Africa," Thurman divines the spiritual desolation of Africans surviving the Middle Passage and addresses the ancestors on the strength of the remembrance of the slave songs:

O my Fathers, what was it like to be stripped of all supports of life save the beating of the heart and the ebb and flow of fetid air in the lungs? In a strange moment, when you suddenly caught your breath, did some intimation from the future give to your spirits a hint of promise? In the darkness did you hear the silent feet of your children beating a melody of freedom to words which you would never know, in a land in which your bones would be warmed again in the depths of the cold earth in which you will sleep unknown, unrealized and alone?[37]

Thurman's point of inquiry into the experience of exile was the question, How does the human spirit accommodate itself to desolation? His essay is one of the most penetrating retrospectives on exilic consciousness ever to appear in African American literature.

Thurman gave deep and honest attention to the ethical issues underlying the struggle for social justice, and he deplored the tendency of the black power advocates and civil rights activists to focus upon methods of dismantling barriers to freedom without fully taking into account what will be demanded of them morally and spiritually once their ends are accomplished:

For those who are engaged in the struggle which is so exhausting and exhaustive, which makes such a primary and absolute demand upon all the resources of one's life, there must be provisions made for some little pocket of energy untapped, ready to move in at the moment of exhaustion when the wall comes down. *This* may be the way, the faith, the givenness of God which will assert itself in our time. And may we find a way to be ready.[38]

Both Thurman and Frazier made use of the symbolism of a wall coming down to describe the desired dismantling of racism in the United States, but clearly the two thinkers had opposite expectations with regard to the role and necessity of religion at that juncture. Frazier projected that the black church would no longer be needed once the system of racial segregation is destroyed, whereas Thurman acknowledged that a prophetic ministry would sorely be needed precisely at the moment when the dismantling effort succeeds.

Indeed, the 1960s and 1970s marked an era of black cultural emancipation in the United States. In those years, the "Africans in exile returned to themselves," an individual and collective process of acknowledging the alienation and victimization of black selfhood in America and the claiming of an ancestral home in Africa. The impetus toward increased black awareness was not exclusively a matter of scholarly discourse; its effects were also felt in the popular culture. Inspired by the television miniseries *Roots,* based on Alex Haley's best selling book of the same title telling of his family's African roots and transition from slavery to freedom in America, many blacks actually made pilgrimages to West Africa in the 1970s, and some became repatriated in African countries. However, the vast majority of American blacks who were attracted to the idea of black power and black consciousness derived their knowledge of Africa secondhand. Instead of a literal return to Africa, they returned to themselves by adopting slogans, hairstyles, and cultural practices associated with the black power movement. Black family reunions became an especially popular way of reclaiming one's roots and acknowledging the strengths and interconnectedness of people of African descent. In the Sanctified churches, Africans in exile "returned" to themselves by constructing vital liturgical and theological frames of reference for making themselves "at home" in America.

In this milieu, that black intellectuals sought to sort out derivative and indigenous cultural forms, both African and American, in a rather complex dialectical process, rejecting selected aspects of American identity and culture, on the one hand, and embracing selected aspects of African identity and culture on the other. West's typology has provided a useful frame for delineating the major options within this process, with the exceptionalist generally valuing the African elements of identity and culture more highly than the American, the assimilationist favoring the American elements over the African, the marginalist employing highly individualistic or idiosyncratic criteria in evaluating the worth of either culture's offerings, and the humanist adopting some scheme of universal values in order to rank and appropriate elements from either culture.

Afro-Pentecostal Thought

Thus far, attention has been given to the contributions of a select group of black intellectuals, writing during the era of black consciousness, whose works referenced in some way the black folk religion of the slave ancestors and/or the urban storefronts. The black cultural trends of the 1970s also elicited a response from a small cadre of intellectuals within the Sanctified church movement whose perspectives have embodied all four categories of the black intellectual tradition. As exceptionalists, these thinkers exalted the black aesthetic of Pentecostal worship; as assimilationists, they embraced the core values of American democratic idealism or of Marxist socialism; as marginalists, they maintained a strong sense of victimization by American racism; and, as humanists, they cherished the cultural authenticity and ecumenical appeal of Pentecostalism. But by virtue of their association with black Pentecostalism, however, they exemplify an additional dimension of exile on the extreme margins of an American society stratified by race, class, and denominational status. This peculiar mode of exilic experience brings to the fore a fifth category of black intellectual exiles, the Afro-Pentecostals.

The Afro-Pentecostal intellectuals of the black consciousness era include Leonard Lovett, James S. Tinney, and Pearl Williams-Jones, all of whom were contributors to *Spirit: A Journal of Issues Incident to Black Pentecostalism,* the premier corpus of Afro-Pentecostal scholarship, which began publication in 1977.[39] Each of these individuals was teaching in a university or seminary during the 1970s while, at the same time, fulfilling a leadership role as pastor or liturgist in one of the Pentecostal denominations. Their scholarly contributions are presented here in summary fashion to illustrate some of the key Afro-Pentecostal exilic perspectives. The Afro-Pentecostal intellectuals represent a significant, if largely unrecognized, scholarly testament to a peculiar configuration of the issues of race, religion, and culture that occurred during the era of black consciousness.

Lovett, a pastor-scholar in the Church of God in Christ, was in the 1970s the first dean of the world's first Pentecostal seminary, the C. H. Mason Theological Seminary at the Interdenominational Theological Center in Atlanta. His 1979 Emory University doctoral dissertation in social ethics explores the relationship between black Holiness-Pentecostalism and black liberation theology. Lovett castigated black Pentecostals who are reluctant to embrace the theory and praxis of black liberation theology. The dissertation's conclusion unites the liberationist praxis of black theology with the mission of the black Holiness-Pentecostal movement:

> In order to engage authentically in its mission to the community of the "hurt," dispossessed and the disenfranchised, the Black holiness-pentecostal movement must be willing to joyfully celebrate the death of rigid ecclesiastical structures and renew itself by involvement in the liberating activity of God.

He had issued an even more stringent demand for the grounding of black liberation in Pentecostal encounter in an article published in the 1973 inaugural edition of the *Journal of the Interdenominational Theological Center:*

Authentic liberation can never occur apart from pentecostal encounter and likewise, authentic pentecostal encounter cannot occur unless liberation becomes the consequence. It is another way of saying no man can experience the fullness of the Spirit and be a racist. This was demonstrated during the early Pentecostal Movement and is evident in the neo-charismatic movement. God was saying to America and the world that there is a Spirit in the world that can bridge racial, denominational and class barriers. If America hears, she can be saved, if the Nation refuses to hear, she will be destroyed from within.[41]

Lovett's jeremiad held America morally and spiritually accountable for maintaining social restrictions based on race, denomination, and class. He specifically condemned the racism of white Pentecostals and charismatics who claim to be possessed of the Spirit but deny the Spirit's unifying imperatives. However, he showed no concern for the gender-based restrictions faced by women, an impression further reinforced by his use of exclusive language in the text.

James S. Tinney taught at Howard University during the 1970s and was one of the nation's foremost curators of black journalism. He was the founding editor of *Spirit*. His 1978 Howard University doctoral dissertation in political science was a comparative study of the Southern Christian Leadership Conference (SCLC), the Black Panther Party, and black Pentecostalism as political and religious movements. In particular, he offered an insightful political reading of black Pentecostal soteriology and drew several general conclusions concerning the political implications and character of the black Pentecostal movement: "That it was born out of revolt against civil religion, that it preserved and embellished practices which were of political import in that they were forbidden practices under slavery and even afterwards. Further, the movement was a manifestation of the nationalist impulse."[42]

Tinney published articles in *Spirit*, in Howard University's *Journal of Religious Thought*, and in a host of anthologies and periodicals. Among his best works is "The Blackness of Pentecostalism," a 1979 article that analyzes Africanisms in Pentecostal religion. In response to the 1970s debate concerning the relevance of Christianity to black people, Tinney argued that Pentecostalism is inherently black. He claimed broad support for such a stance, by implication, from scholars who refer to Pentecostalism as "a third force," "the religion of the poor and dispossessed," "primitive religion," or as a religion common to the "culture of poverty." His interpretation of the 1906 Azusa Street revival was that "Africanisms in worship, long suppressed by slaves and hidden from white view by captive Black people, suddenly came out into the open."[43] Tinney's exceptionalist perspective on the essential Africanness of Pentecostalism was presented with an interesting twist; he claimed that Pentecostalism, "the only new Black religious form to arise indigenous to this country," imposed its distinctive Afro-American forms onto several million whites but was not regarded as an entirely new phenomenon in Africa because of the prevalence of Pentecostal practices in African traditional religions and independent churches prior to the revival. The article ends with the assertion that Pentecostalism is radically social and political; as such, it entails artistic rebellion against the "humiliating deadness of Western culture," rejects white cultural values, and bears a power theme that has been translated into political rebellion, especially in the developing nations of the world: "Regardless of the channel some would like for the revolution to take, Pentecostalism (outside of white America) shows no affinity for capitalism."[44]

In 1980 Tinney organized the Pentecostal Coalition for Human Rights, a support group for racial and sexual minorities. He was the founding pastor of Faith Temple, a Pentecostal congregation in Washington, D.C., that welcomed gay and lesbian members. In view of his strong advocacy for gay and lesbian issues in the church, it is unfortunate that his work has been virtually ignored by black, feminist, and womanist liberation theologians who are addressing issues of sexual preference in their writings. Tinney died of AIDS in 1988 at the age of 46.[45]

Pearl Williams-Jones was a professor of music at the University of the District of Columbia and a musician at the Bible Way Temple in Washington, the mother church of the denomination headed by her father, Bishop Smallwood E. Williams. She was an accomplished pianist, vocalist, and artistic consultant whose repertoire included the European classical tradition, the Negro spirituals, and gospel music. Williams-Jones pioneered the inclusion of gospel music in public school and university curricula and worked closely with the Smithsonian Institution to document, preserve, and present public performances of black folk music traditions. At the time of her death from cancer in 1991, she was musical director of a gospel concert series sponsored by the Washington Performing Arts Society at the Kennedy Center for the Performing Arts in Washington.

Most of Williams-Jones's published works deal with black sacred music and folk religion. She thoroughly described and analyzed the black aesthetic in gospel music in a 1975 article written for the journal *Ethnomusicology*. The notion of a black aesthetic was an important issue in the black consciousness movement. In terms reflective of the times, she defined black gospel music as "a declaration of black selfhood" and an "underground or counterculture body of music" that is resistant to assimilation: "Black gospel music has not consciously sought the assimilation of European religious music practices or materials into its genre. If this has occurred, the materials have been improvisationally recreated to conform to black aesthetic requirements of performance."[46] Williams-Jones viewed gospel music as the dominant force in the preservation of black cultural identity and as the "positively crystallizing element in the emerging black aesthetic."[47]

Williams-Jones's study of women in the Pentecostal churches was based upon her extensive firsthand knowledge of the leading women in the Sanctified church movement. She stands alone among Afro-Pentecostal intellectuals of the black consciousness era in her appreciation of the relevance of women's roles and perspectives to the overall struggle for human rights and black cultural emancipation.[48]

In general, the black Protestant churches supported the civil rights movement on many levels—providing leaders for the movement from the ranks of their clergy and laity; promoting mass involvement via networks of local congregations, denominations, and church-based organizations; and proffering songs, prayers, and sermons from the black folk tradition to inspire participants in the demonstrations, marches, and rallies.[49] Black Pentecostals participated in these efforts but were also directly involved in two events that framed the transition of black America into the civil rights movement and subsequently into the era of black consciousness.

The first such event is the lynching of Emmett Till, the Chicago teenager who allegedly had whistled at a white woman while he was visiting relatives in Mississippi during the summer of 1955. The Till case galvanized the black demand for racial justice in 1955. Sociologist Cheryl Townsend Gilkes has noted that the boy's mother, Mamie Till Mobley, was a member of the Church of God in Christ. It was she who insisted on the release and return of her son's mutilated body for a public wake at a Chicago funeral home, where nearly a quarter of a million people filed by the open casket. Gilkes cited a black press account stating that "four months before Rosa Parks took the personal stand against segregation, a Black Chicago mother, Mrs. Mamie Till Mobley unknowingly but decisively jolted 'the sleeping giant of Black people.' "[50]

The second event is Martin Luther King Jr.'s sermon proclaiming his vision of the "promised land" and foretelling his own demise. It was delivered on April 3, 1968, the eve of his assassination, at Mason Temple, world headquarters of the Church of God in Christ, at a rally in support of sanitation workers in Memphis. Lovett has commented that this occurrence was not coincidental or accidental, given the church's historic identification with the voiceless, downtrodden, and the oppressed.[51]

7

The Church in Exile: Vital Signs
Outside the Mainstream

Summary: The Dialectics of Exilic Existence

The Sanctified church tradition is an African American Christian reform movement that seeks to bring its standards of worship, personal morality, and sociocultural concern into conformity with principles of holiness and spiritual empowerment. All groups within this rubric adhere to the doctrine and practice of sanctification in some form and historically have embraced an ascetic ethic forbidding alcohol, tobacco, other addictive substances, gambling, secular dancing, and immodest apparel. The Sanctified churches are congregations of saints, an ethical designation members apply to themselves as an indication of their collective response to the call to holiness. The saints follow the holiness mandate in worship, in personal morality, and in society, based upon a dialectical, exilic identity of being "in the world, but not of it." These saints are fully aware of their marginalized status, based upon racial and religious differences, within the dominant culture; thus, exile has been offered as an appropriate paradigm for interpreting their experience.

Throughout this study, the self-understanding of the adherents of the Sanctified church movement has been emphasized and expressed in terms of a complex of "both–and" dialectics corresponding to various aspects of the exilic ethos. The first and most compelling of these dialectics is the situation of being "in the world, but not of it," as described in the introduction with reference to the historical marginalization of the Sanctified church movement in light of its African antecedents and its reformulations of slave religion. The discussion of the Sanctified church as a response to the problem of race, sex, and class in American Protestantism in the first chapter reveals a dialectics of egalitarian and exclusivist notions of who is included in the body of Christ. In the second chapter, which profiles a contemporary Holiness congregation in Washington, D.C., the dialectics of refuge and reconciliation characterizes the complementary themes represented in its pastoral leadership and corporate ministry. The third chapter develops a dialectics of static and ecstatic elements in order to analyze worship in the Sanctified tradition. An attempt is made in the fourth chapter to shed new light on the dialectics of the sacred and the secular with reference to the trajectories of religious and popular culture that comprise gospel music. The dialectics of protest and cooperation within educational and religious institutions is discussed in the fifth chapter with respect to the collegiate gospel choirs and the black clergy caucuses that grew out of the black consciousness era of the 1960s and 1970s. In the sixth chapter, which reviews four categories of intellectual approaches to the study of black folk religion during the same period, the key dialectics emerging from both the studies and the times is the tension between African and American identity.

This complex of exilic dialectics—in or not of the world, egalitarian or exclusivist norms within the church, refuge or reconciliation ministry styles, static or ecstatic worship, sacred or secular music, protest or cooperation in institutional life, African or American identity—although not intended to be exhaustive, is hoped to represent a progressive step beyond the "double-consciousness" described by W. E. B. Du Bois in 1903, which persists as the dominant paradigm in African American religious and cultural thought: "One ever feels his twoness—an American, a Negro; two souls, two thoughts, two unreconciled strivings."[1] Many contemporary scholars cite Du Bois's notion of "double-consciousness" as foundational for their own reflection, among them, Riggins Earl, Paul Gilroy, Michael Harris, C. Eric Lincoln, Lawrence H. Mamiya, Peter Paris, Theophus Smith, Emilie Townes, Theodore Walker Jr., William Watley, and Cornel West.[2] Only West brought a critical response to bear on Du Bois's characterization of African American existence, albeit somewhat vicariously through the mind of Malcolm X:

> For Malcolm X this "double-consciousness" pertains more to those black people who live "betwixt and between" the black and white worlds— traversing the borders between them yet never settled in either. Hence, they crave peer acceptance in both, receive genuine approval from neither, yet persist in viewing themselves through the lenses of the dominant white society. For Malcolm X, this "double-consciousness" is less a description of a necessary mode of being in America than a particular kind of colonized mind-set of a special group in black America. Du Bois's "double-consciousness" seems to lock black people into the quest for white approval and disappointment owing mainly to white racist assessment.[3]

In his 1970 doctoral dissertation on Francis J. Grimke, church historian Henry Ferry presented a similar social ethical critique of Du Bois's view through the perceptual lens of the "Black Puritan":

Nevertheless, it seems that Du Bois has been misled. Perhaps the dichotomy he found in the soul of black folk is but the deceptive reflection of the radical division in the heart of white men described by Myrdal in the *American Dilemma*. The real antithesis, therefore, is not Negro/American but American/hypocritical American. How else does one account for the fact that Grimke developed his critique of the nation in terms of its own principles?[4]

Ferry's statement makes feasible the separation of the question of what ongoing significance racial identity has for African Americans from moral inquiry into the problem of white racism in America. On this basis, the black American's social double-consciousness is a burden imposed by the moral double-consciousness or duplicity of the white American whose self-understanding and social behavior are guided by racism.

Two of the most influential books on the black church from the standpoint of methodology, Peter Paris's *The Social Teaching of the Black Churches* and Lincoln and Mamiya's *The Black Church in the African American Experience*, each fashion a complete dialectical analysis of black church data on the basis of Du Bois's "double-consciousness" but without critique. It is hoped that the present study contributes to the ongoing quest for robust methodological approaches to the study of black religion by demonstrating that a dialectical interpretive paradigm can be developed from the sources themselves, one that fully acknowledges the complexity and integrity of the tradition from the vantage point of the believers. What seems especially important here is the realization that the "both–and" approach to dialectical analysis differs significantly from the "either-or" perspective, in that the former is characteristic of individuals and groups (such as those described in this study) who can find equilibrium and fulfillment between extremes, whereas adherents to the latter *either* demand resolution or suffer greatly in the tension, as is the case with Du Bois's description of the agony of "double-consciousness," as "two warring ideals in one dark body, whose dogged strength alone keeps it from being torn asunder."[5] The exilic concept allows for honest appraisal of and response to white racism without, at the same time, having one's own identity totally shaped by it; the dialectical understanding of existence need not become totally collapsed under the weight of oppression.

Toward an Exilic Ecclesiology

Most black scholars who have found intellectual refuge in Du Bois's paradigm have also embraced liberation theology as a theological method and ecclesiological perspective. Some of the Afro-Pentecostal scholars, notably Leonard Lovett and James Tinney, have employed liberation theology and black theology in their own work. Among the most enlightening recent challenges to liberation theology, one that speaks specifically to some of the concerns implicit in this study of the Sanctified church tradition, is a short article written by a doctoral student for the *Christian Century:* "From 'Liberation' to 'Exile': A New Image for Church Mission." The author, Ephraim Radner, a white Episcopal priest, outlined an "exilic ecclesiology" based on insights he gained while working in an urban parish in Cleveland. His proposal was directed to his own context, the churches of the white Protestant mainstream, but his allusions and applications related as well to the Sanctified churches on the margins.

Radner began with a critique of the inadequacies of liberation theology based on his own experience as a "socially responsible" seminarian in the 1970s who concluded that "liberation theology was the only show in town": "Those of us who were uneasy with the simplistic and politically rote conclusions to which liberation theology gave rise had no other examples to follow. Tired North American denominations offered no parallel to the martyrs and confessors of the church in Latin America, Africa and Asia."[6] He observed that liberation themes dominate church presses and seminaries; it is the "almost normative lens" through which churches are asked to see their mission. Yet, in reality, the same churches who address their liberation rhetoric in the form of statements, resolutions, protests, pickets, and boycotts have failed to engage the poor directly and have little to offer with respect to the practical task of reforming communities. Meanwhile, the poor are excluded from liberal and conservative denominational churches, both black and white:

> Statistics indicate that major denominations—including traditional black churches—have increasingly exclusive memberships, defined along economic, class and ethnic lines. Churches in inner cities and poor rural areas are closing, while those that remain are often composed of commuting members with little interest in the church's neighbors.[7]

Citing an alarming example of rhetorical posturing from a Christian social justice magazine, Radner showed how a liberationist outlook obscures rather than clarifies the practical imperatives of Christian ministry.

Radner offered a threefold remedy for churches seeking to engage the poor directly as they address issues of crime, drug use, and housing. First, he called for "totalistic transformation." Behaviors, expectations, attitudes, and motivations must be addressed in conjunction with any attempt at fostering social change. Beyond providing shelter for the homeless, it is necessary to offer services, training, support groups, monitoring, and sanctions on miscreant behavior. He argued that liberationists seek to deliver people from the contexts that foster crime and drug use but are unwilling to engage in the totalistic social intervention needed to break the bonds that create such contexts; they issue prophetic calls devoid of the practical guidelines their commitment to political change would entail. Nevertheless, the church has traditional avenues for faithful response to crime and drug use because totalistic renewal is the voluntary province of the church in the United States The groups who are engaged in totalistic renewal tend to be black Pentecostals and dissenting Protestants:

> Reaching out and incorporating new members and inculcating in them behaviors and expectations congruent with the vision of the Christian community has been the main activity of the churches' mission across the centuries. It has been the means for the transformation of many socially marginal groups in the U.S., from poor rural whites in Methodist and Assemblies of God churches to rural and dislocated urban blacks in Baptist and Church of God in Christ churches. These highly self-conscious groups have deliberately, and on the basis of well-articulated evangelical mandates, provided the means for motivational and educational upbuilding. In so doing, they have also been the vehicles for upward mobility.[8]

Radner named the secular culture of economic consumerism as a major factor in the churches' abandonment of their traditional mission of totalistic renewal. Only the "sects" such as Jehovah's Witnesses and the Black Muslims have maintained the vision and motivation to seek and transform those outside the economic and educational mainstream. These groups are vilified by the political right, on the one hand, for refusing to accept patriotic values and for resisting participation in the political process, and by the political left, on the other, for uncritical espousal of economic self-improvement and failure to challenge the capitalist system. However, these critiques hint at the coherence and integrity of these groups as separate communities capable of totalistic renewal.

The notion of separation leads to the second aspect of Radner's remedy, the idea of the church as an alternative community with an identity of "exile." In this regard, he put his finger on the fundamental ecclesiological fallacy of liberation theology, that in their zeal to merge the sacred and secular, liberationists reject the notion of the church as a separate community that evangelizes the people it seeks to serve:

> Liberationism fails to liberate today because of a general inability to understand the church as a community apart from secular society—a separate community with both the clearly defined means and the vocation to evangelize and form new members from among those whom the society at large has

abandoned. It makes little sense for churches to work at becoming liberating agents when they refuse to become communities that are themselves inclusive of the people they seek to serve.[9]

Radner argued that liberation also has failed as a theme for church mission insofar as it denies the sectarian nature of reform built into the nation's polity: "It is rather the growth and expansion of religious communities, *separate* but *within* the larger society, that will engender vehicles for noncoercive deliverance."[10]

Exile is offered as a more effective theme for the church's mission than liberation. A church with an exilic identity is an alternative community formed by a coherent set of values that are at odds with the surrounding culture. Again, Radner looked to history and tradition to support his point of view:

> Traditionally the Christian church has internalized its own experience of exile within an alien culture and adopted a spirituality suffused with exilic qualities. For Christians exile has been not only a condition forced upon a small group of people but a state into which everyone was called by God for their human maturation—a place of formation, where attitudes and motivations were molded by a community without earthly roots.[11]

The Puritans struggled with the concepts of exodus and exile; the nineteenth-century evangelicals are offered as examples of a truly exilic perspective linked to missionary outreach that expanded the community of churches to include socially marginal groups (their effectiveness among blacks being a major historical point of departure for the study of the Sanctified churches, as indicated in the introduction). Radner stated that the notion of exile is poorly understood by Americans, but it is not at all clear that he was including African Americans in that assessment.

The third and final element of Radner's proposal was a return to New Testament ecclesiology in support of the exilic identity of the church's mission: "Much of the New Testament calls people *out* of an existing society into a theocratic alternative that is to continue in the midst of the larger society until the inbreaking of God's own action."[12] He lifted up the First Epistle of Peter as the best New Testament example of exilic ecclesiology but did not offer any specific texts or details, simply remarking that God's elective love buttresses the more practical elements of the church's posture. This epistle is addressed to the "exiles of the Dispersion" (1 Pet. 1:1b, NRSV), and in each of its five chapters texts that address the exilic identity of the church have a special appeal in the Sanctified tradition: "be holy, for I am holy" (1:16b, NRSV); "ye are a chosen generation, a royal priesthood, an holy nation, a peculiar people" (2:9a, KJV); "sanctify the

Lord God in your hearts" (3:15a, KJV); "Above all, maintain constant love for one another" (4:8a, NRSV); "Be sober, be vigilant; because your adversary the devil, as a roaring lion, walketh about, seeking whom he may devour" (5:8, KJV).

A distinctive Christian identity involves a host of special commitments and "hard confrontation" with secular values, pointing the church to the way of the cross as the social character of exile. By maintaining its integrity in the face of the dominant culture, however, the church can rediscover the means to embrace new members from the margins and to form a distinct and enticing place of renewal: "Exile is a move to that realm where divine liberation can begin to take on meaning, because it springs from the longing of a separated people. If the churches cannot groan, God shall never hear. That is the secret of the exodus, and the moment to which we are called to return."[13] Certainly, the "longing of a separated people" is implicit in the sacred song, speech, and dance of the Sanctified churches, whose worship, on this view, represents a more authentic emulation of divine liberation than the many conferences and meetings convened by liberationists to reflect with great eloquence on the plight of the poor, in settings where the poor are not welcome and prayer is out of order. At least one liberation theologian, Harvey Cox, has admitted that Pentecostalism is potentially more radical and revolutionary than liberation theology:

> I believe that Pentecostalism, and the global upsurge of spirituality it represents, may in the long run have a considerably more radical, even revolutionary, impact than liberation theology can. At its best, Pentecostalism attacks not only the demonic political and economic systems that keep God's children in cruel bondage, but the core of distorted values and misshapen worldviews that sustains these oppressive structures.[14]

The holiness dialectics of being "in the world, but not of it" has formed religious communities that are *separate* but *within* the larger society. Radner's proposals for totalistic transformation, exilic ecclesiology, and reiteration of the biblical call to "come out" to form alternative communities that welcome the poor and the marginalized both confirm and challenge the self-understanding of the Sanctified church tradition. Totalistic transformation is precisely the agenda of sanctification and deliverance. Exilic ecclesiology exactly expresses the viewpoint and experience of those who have come out or been forced out of mainstream Christian communions because of racism or denominationalism. Sanctified storefronts and other edifices stand as places of refuge and reconciliation in the inner city neighborhoods long abandoned by affluent white and black Christians. At the same time, it must be recognized that small urban congregations of saints that aggressively evangelize the poor typically lack access to the facilities, funding, and bureaucratic support of the more affluent mainstream denominations. Third Street Church of God, for example, has yet to marshal adequate human and fiscal resources to provide the full complement of services, monitoring, and support needed by the homeless people who worship there daily.

Those Sanctified churches that have succeeded as vehicles of upward mobility need to be challenged to decide whether the goal of ministry is spiritual formation or socialization to the middle class. Hans Baer and Merrill Singer identified within the black Holiness-Pentecostal movement a growing shift away from traditional concerns with spiritual salvation to an acceptance of bourgeois values of temporal success and material acquisition; James Tinney has offered an even more pointed critique of the materialistic excesses of black Pentecostals.[15] The distinctive Christian lifestyle that involves confrontation with the values of the secular environment is properly associated with the way of the cross, but the Sanctified churches are subject to the same forces of secularization and accommodation noted by Finke and Stark in *The Churching of America* (following H. Richard Niebuhr's *The Social Sources of Denominationalism*), which tend to pressure religious bodies to move from a high state of tension with their environments to a lower state. The typical response to such pressure is to relax moral standards; the typical result is the loss of members and viability, a condition that plagues the churches of the Protestant mainstream.[16] Moreover, a congregation that espouses an ascetic lifestyle and forbids consumption of alcohol and drugs is in some ways better positioned to relate the complex behavioral issues involved in crime and drug use to the evangelistic message directed to the urban poor.

The Nation of Islam has made great inroads in winning converts among the urban poor and the prison population, precisely because of its bold invitation to structured spiritual formation, a strict asceticism not unlike the Sanctified ethic, and totalistic life change. Ironically, many Christian groups who criticize the Muslim approach have no compelling message, no effective methods, and no attractive models of their own to offer in its place and are in no way motivated to evangelize the poor and incarcerated black men targeted by the Muslims.

In the course of his commentary on the historical relationship between the black struggle for racial equality and the ascetic ethics of the social gospel, Wilson Moses lamented the rejection of both by modern secularists and religious conservatives alike:

> The present hostility towards the values of a social gospel does not bode well
> for the struggle for racial equality. The battle for the rights of black Ameri-

cans has traditionally been a part of the campaign for moral reform. It was in Boston's churches, not in New Orleans' brothels, that reformers linked together temperance, antislavery, and the struggle for women's rights before the Civil War. The key figures in the American civil-rights struggle have all borne the mark of puritan discipline and self-denial. This is not to suggest that either America's black population or its leaders may be characterized as "puritanical." It is simply to illustrate that the black liberation tradition has never been libertine.[17]

In Moses's view, the black liberation tradition has been inextricably interwoven with the American traditions of evangelical reform, perfectionism, and the social gospel and bound up historically with the quest for moral excellence and discipline. His characterization of black leaders in America who preach conservatism in lifestyles and liberalism in politics is perhaps more directly reflective of the posture assumed by the Sanctified churches than of the positions taken by the traditional black denominational leadership in this century. It remains to be seen whether the present generation of civil rights leaders and organizations will continue their current emphasis on the problems of truancy, crime, drugs, and sexual activity among black youth from the vantage point of advocating black family values. It is not clear how such an impetus can be sustained apart from the formation of communities in which a coherent ethic of self-discipline and delayed gratification is promoted, rewarded, and embodied by adults and youth together.

In a second article published in the *Christian Century,* "Religious Schooling as Inner-City Ministry," Radner has lifted up the Grace Christian Educational Center in Brooklyn, New York, as an example of what "sectarian" Pentecostal churches can do to offer "communal, cultural, and moral formation to children cast adrift in violence and hopelessness." He saw this form of formative evangelism as possibly the greatest contribution churches can make to the mission of Christ in the cities:

> Church schooling—that is, religiously sectarian schooling—has proved successful thus far in educating poor urban children of many ethnic backgrounds. In addition to practicing those methods that researchers say are necessary for success, Christian church schools possess the knowledge of the gospel of Christ, the blessing that derives from living in this knowledge, the power that is embodied in this life and the joy that comes from sharing it with others. This is what we used to call "mission."[18]

The urban church school could potentially become the catalyst for a new flowering of cooperative ventures in urban mission in which mainstream Protestant and Holiness-Pentecostal churches join forces with middle-class black professionals to meet the spiritual and educational needs of poor children in the inner city. The U.N.I.Q.U.E. Learning Center described in chapter 2 is not a school, but it is a replicable model for educating urban youth with limited human and fiscal resources. Tutoring, learning experiences, and counseling are provided for children and teens after school and during the summer, based on creative partnerships with the Third Street Church of God and other churches and organizations that bring these youth into regular contact with adults who care.

The Sanctified church is a church in exile; the tradition comprises a family of African American religious communities that exemplify the exilic ecclesiology of the New Testament. Here a summary statement of three distinctive features of Sanctified church ecclesiology in exilic perspective—ethics, spirituality, and evangelism—attempts to signify what congregations inside and outside the tradition might do to enhance their position as viable centers of Christian formation and totalistic transformation that engage in meaningful ministry and worship, inclusive of all who desire to respond to the call to holiness, irrespective of sex, race, or economic status.

The Ethics of Holiness and Unity

Ethics has been ascribed primary significance in the basic definition of the Sanctified church tradition. Most scholars and casual observers of these churches have noted the emphasis on personal morality and ascetic lifestyles in this regard and the corresponding prohibitions against alcohol, tobacco, addictive drugs, extramarital sex, gambling, secular dancing, and the like. In the minds of observers and participants alike, this attention to personal morality has sometimes obscured the focus on other dimensions of holiness ethics, as appropriated from biblical sources. The Holiness tradition includes a three-dimensional holiness ethic, mandating not only holy living for individuals but also holy worship in the churches and holy justice in the social order. The biblical mandate to holiness can be seen as a composite of three distinct Old Testament traditions:

> Diversity within unity is to be discerned in the fact that for the different groups of religious persons within Israel—prophets, priests, and sages—the kind of cleanness required by holiness varied. For the prophets it was a cleanness of social justice, for the priests a cleanness of proper ritual and maintenance of separation, for the sages it was a cleanness of inner integrity and individual moral acts.[19]

Thus the call to holiness is not simply an admonition to stay sober and celibate; it is a vocation to bring personal lifestyle, corporate worship, and social engagement into harmony with the attributes and demands of a holy God. In the Sanctified church tradition, the possessing Spirit is the Holy Spirit, the pursuit of social justice is a holy mandate, and the purity of the saint is a testament to God's holiness.

In light of a New Testament understanding of holiness, William C. Turner Jr. has challenged Pentecostals to revisit their holiness roots in his 1978 article "Is Pentecostalism Truly Christian?" He endeavored to harmonize holiness ethics and Pentecostal empowerment by means of the proposition that "holiness is the appropriation in living of the power in the blessing that is the Pentecostal experience."[20] Citing the ethical dilemma of the spiritually gifted Corinthian church in which love was sorely lacking, he suggested that the biblical love ethic be given priority by today's Pentecostals:

> In Corinth, for example, there was no indication of any lacking in spiritual gifts. Yet the people did not know how to love one another; and God is love. Love is the first and most basic fruit of the Spirit of Holiness. The blessings of God can come quite apart from his righteousness and holiness. This is why people can shout, prophesy, speak in tongues, and dance all night, and get in a lot of devilment the next day.[21]

Turner exhorted Pentecostals to "take time to be holy" and to learn to appropriate the virtue of their spiritual experiences in lives of holiness in which the Spirit of God lives out the life of Jesus Christ to whom the gospel story bears witness.

The question of how to achieve a practical and productive balance between the ecstatic gifts and the static fruits of the Spirit remains the critical ethical ecclesial challenge facing the Sanctified church movement today. The Holiness "camp" tends to discredit the gifts, and the Pentecostal "camp" sometimes exalts them above all else as criteria for participation and witness. To complicate matters further, even when a charismatic movement emerges within the Holiness church, there is a tendency to emphasize gifts over fruit. Specifically, attitudes toward tongues-speaking have occasioned a great deal of controversy and division in the Holiness-Pentecostal movement, a situation that seemingly can be resolved only through mutual commitment to balanced approaches that are respectful of persons given to ecstatic praise as well as to those dedicated to ethical rigorism.

The holiness ethic as articulated by the Church of God reformation movement suggests a three-dimensional notion of unity, grounded in the refutation of sexism, racism, and elitism as barriers to oneness in the body of Christ. Although many glaring inconsistencies and failures are cited in this study and elsewhere in this regard,[22] it is also the case that William J. Seymour, the Brothers and Sisters of Love, and other pioneering witnesses in the Sanctified tradition preached and practiced unity to great effect. Seymour's orientation to the doctrines of holiness and unity espoused by the Church of God prepared him for the interracial, international outreach across gender lines that characterized his leadership of the Azusa Street Revival and the Apostolic Faith Mission. Elisha and Priscilla Wimbish and their followers were convinced that God had directed them to establish the campground at West Middlesex as a haven for saints who would experience ostracism and discrimination in churches that embraced the dominant racist, sexist, and elitist mores of the times.

Race has been a constant source of Christian disunity and discord in U.S. church history. Arguably, Christian racism is the predominant factor that has shaped denominational life in North American Protestant churches, many of whom split during the antebellum period over the slavery issue. Martin Luther King Jr.'s comment that eleven o'clock on Sunday morning is America's most segregated hour stands as an indictment of Christian denominational racism, which seems virtually unaffected by the civil rights struggle and the removal of legal barriers to racial integration in other social institutions. Holiness minister and author Frank Bartleman was so impressed with the racial integration at Azusa Street that he exclaimed that "the color line was washed away in the blood,"[23] but racism eventually overwhelmed the Pentecostal movement, and racial barriers were systematically reinstated. Racism is the premier social context for formation of the exilic consciousness of African Americans because all have been victimized in some way by racial discrimination, regardless of religious affiliation. By the same token, Christianity has generally failed to dismantle and disarm the white racists within its own ranks, again irrespective of denomination.

As Cheryl Townsend Gilkes and Pearl Williams-Jones have noted, black women have struggled for prominence in the Sanctified churches; nevertheless, the tradition has been much more open to and accepting of women's gifts and leadership than the white and black denominational churches. This conclusion is supported by a 1986 study showing that the majority of ordained women clergy in the United States are in the Holiness-Pentecostal churches and the minority are from the Protestant mainstream, a disproportion that demonstrates the extent to which the Protestant denominations, even the liberal ones, lag behind the Holiness-Pentecostal movement in the validation of women as equals.[24] The majority of worshipers in the Holiness-Pentecostal churches are women, even in those communions that deny women equal status.

Class differences present barriers to Christian unity that are as formidable as those imposed by race and sex. However, the poor have been readily accepted into the Sanctified churches, seen by many as the churches of the disinherited, and have been less welcome in the more affluent white and black churches of the Protestant mainstream. The label "storefront church" projects the image of a small group of saints meeting in a rented space in an inner city neighborhood in which the poor have an open invitation to worship.

One of the best studies to date of the outreach of the Holiness movement to the poor is Norris Magnuson's *Salvation in the Slums: Evangelical Social Work, 1865–1920.* He told the story of the "gospel welfare movement," whose openness toward and identification with the poor was based on continuing close personal contacts and settlement houses whose workers literally became "neighbors of the poor." The best-known group within this movement is the Salvation Army, which is identified with the Wesleyan-Holiness tradition. Although participants in the gospel welfare movement centered their faith in the ethical principle of love, rather than in a doctrinal creed, Magnuson acknowledged, their constant practical emphasis helped them keep that principle from evaporating into sentimentality: "The right kind of religion is love with its coat off, doing its best to help somebody."[25] The love ethic empowered this movement to embrace as worthy and capable persons blacks, Asians, and other groups the social gospelers and leaders of the larger progressive movement appear to have neglected: "In a day when the plight of blacks was worsening, the Salvation Army and kindred organizations defended them and welcomed them into rescue institutions for assistance and as fellow-workers."[26] One of the main criticisms directed toward the gospel welfare movement and its modern-day successors is failure to give adequate attention to the pursuit of justice. Magnuson cited evidence that the leadership fully understood the poverty they encountered in its social, political, and economic context:

> "We must have justice—more justice," the *New York Tribune* quoted Ballington Booth, commander of the American Salvation Army forces, as saying. "To right the social wrong by charity," he had said, "is like bailing the ocean with a thimble. . . . We must readjust our social machinery so that the producers of wealth become also owners of wealth."[27]

The Salvation Army and similar groups have been unfairly disparaged as a hindrance to social Christianity, mainly due to their insistence on evangelizing the poor. Magnuson argued that the "revivalistic and holiness faith of these people produced extensive social programs and close identification with the needy."[28] The urban prayer breakfast at the Third Street Church of God offers a contemporary example of ministry to and among the urban poor, motivated by essentially the same ethic and evangelistic concern described by Magnuson. It is a model of three-dimensional Christian unity among persons of both sexes, all races, and differing economic and educational status coming together regularly for worship, sharing, and a hearty communal meal.

Spirituality and Christian Formation

Homileticians Ella P. Mitchell and Henry H. Mitchell have defined spirituality as "sensitivity or attachment to religious values," involving a belief system about God and creation that "controls ethical choices/behavior and supports calmness of spirit in times of stress." Black spirituality adds the testimony that God as Holy Spirit is real and present with persons.[29] The promotion and cultivation of spirituality is a central function of worship in the Sanctified church tradition; to refer to one another as "saints" denotes devotion to a particular constellation of spiritual beliefs, practices, and effects. Much has already been said in this study concerning the nature of spirituality in the Sanctified churches without necessarily naming it as such. Robert Franklin, a Church of God in Christ clergyman and ethicist, has developed a general typology for African American spirituality that is employed here as a guide to developing a more explicit paradigmatic understanding of Sanctified spirituality. In his presentation, a type, theology, praxis, and example are offered for each of six traditions of African American spirituality.[30] His examples are here set forth and in some cases augmented to include noteworthy representatives from the Sanctified church tradition.

First is the contemplative tradition, which lifts up intimacy with God as the central virtue and value in Christian formation. Its praxis is prayer and meditation. The example is Howard Thurman; James Earl Massey also exemplifies this tradition.

Second is the holiness tradition, which stresses purity of life and thought achieved by means of practices such as fasting, prayer, and renunciation of the world. Exemplary of this tradition is Mother Lillian Brooks Coffey, head of the Church of God in Christ Women's Department and "right hand" to Bishop C. H. Mason.

Third is the charismatic tradition, emphasizing personal empowerment through the Spirit. Its characteristic discipline is tarrying, a Pentecostal method of seeking the outpouring of the Spirit in worship. The example is William J. Seymour, shepherd of the Los Angeles Azusa Street Revival.

The social justice tradition, fourth, pursues public righteousness, the moral hygiene of the entire society, and the protection of the poor. Its praxis is public activism and community ministry; its exemplars are Martin Luther King Jr. and his pastoral predecessor at Dexter Avenue Baptist Church, Vernon Johns. There are many others worthy of mention from the Sanctified church tradition: Bishop C. H. Mason, founder of the Church of God in Christ who was imprisoned for taking a pacifist stance in opposition to U.S. involvement in World War I; Bishop Arthur Brazier, organizer of the Woodlawn Organization for low-income blacks in Chicago; the Reverend Willie Barrow, former national director of Operation PUSH; the Reverend Addie Wyatt, retired international vice president and director of civil rights and women's affairs for the United Food and Commercial Workers International Union, who is copastor of the Vernon Park Church of God in Chicago with her husband, the Reverend Claude Wyatt Jr.; and Bishop Herbert Daughtry, head of the House of the Lord Pentecostal Church in Brooklyn, president of the African Peoples' Christian Organization, and socialist organizer.

The Afrocentric tradition highlights the cultural distinctiveness and aesthetics of African people through the recovery and retrieval of drums, guitars, tambourines, and African modes of worship. The example given is African Methodist Episcopal Bishop Henry McNeal Turner, who strongly advocated emigration to Africa during the period between Reconstruction and World War I. Another example is William Christian, a former slave who founded the Church of the Living God (Christian Workers for Fellowship) in 1889 and advocated the belief that Jesus and the saints of the Bible were black. This tradition is exemplified today by African Methodist Episcopal Bishop John Bryant, who, in partnership with his wife, the Reverend Cecelia Bryant, and several pastoral protégés, has initiated a "Neo-Pentecostal" movement in his denomination that embraces African liturgical practices and symbols. The Pentecostal heritage of the principal architect of the Afrocentric idea, Molefi Asante, has undoubtedly influenced his thought and approaches. Although the concept has not been forthrightly embraced by saints—who would prefer to be described as "Christocentric"—the fact remains that the Sanctified churches are more obviously expressive of Africanness than any other institution in black life.

Finally, the evangelical tradition stresses knowledge and study of God's word. Its principal disciplines are proclamation and teaching. The example is religious broadcaster Frederick K. Price. Another is Elder Lightfoot Solomon Michaux, founder of the Gospel Spreading Church of God and pioneer in worldwide media-supported evangelism, whose *Happy Am I* evangelistic broadcast attracted an estimated weekly audience of 25 million listeners on the CBS Radio Network during the 1930s.[31]

Franklin concluded that the richest African American congregations embody all six traditions. Exemplars for each tradition can readily be identified from the Sanctified churches, indicating that these congregations are well positioned to integrate the six dimensions of African American spirituality and Christian formation.

Biblical and Ecumenical Witnesses of the Gospel Message

The saints are a people of the book, who claim the Bible as text for their testimony of being saved, sanctified, and filled with the Holy Ghost. In his best-selling 1993 book, *The Culture of Disbelief,* Stephen Carter has cast black American Christians in a fundamentalist, socially conservative mode, on the basis that they are much more likely than other Americans to treat the Bible as literal truth, are "heavily represented" in the nation's more conservative Protestant denominations, and are among the most likely to support a return to traditional roles for women.[32] Obviously, Carter's formulation disregards black Holiness and Pentecostal Christians. Biblical interpreters in the Sanctified church tradition are not often fundamentalists; they are perhaps best described as liberal literalists. They are liberal because the exilic experience of being black in a racist society forbids them to follow uncritically conservative, fundamentalist readings fostered by the descendants of those who used the Bible to justify slavery. Their collective history and experience make it unfeasible for them to engage in conservative political and social behaviors characteristic of those who have a stake in maintaining the status quo. They are *literalists* because they are unwilling to surrender the present authority and power of the Scriptures to the forces of modernity and are generally resistant to liberal readings fostered by modern black and white exegetes. Since the days of the "proto-Pentecostal" and "proto-Holiness" evangelists like Julia Foote and Amanda Berry Smith, who encountered racism, sexism, and elitism as they endeavored to preach the gospel, the saints have discerned within the Scriptures a literal justification for liberative theology and praxis. These liberal literalists refuse to be held hostage to the historical-critical method.

New Testament scholar Vincent Wimbush has addressed some of these concerns in a series of writings that evaluate black approaches to the Bible. He categorized the biblical views of blacks in the Holiness-Pentecostal churches as esoteric readings of a type shared in common with other "sects" such as Black Muslims and Black Jews. His description of religious communities reflects the exilic perspective:

> With a more critical perspective of the world and of American society and its biblical self-understanding, these groups are different from the worldly and mainstream Baptists and Methodists, among others. . . . In sum, such groups can be characterized by their consistent rejection of both American society in general and the older established African American religious communities. The former is rejected on account of its racism; the latter are rejected on account of their accommodationism.[33]

He saw as ironic the ability of these "separatist" groups to achieve the integration that eludes black and white mainstream religious communities; through the esoteric books and esoteric knowledge about such books, a new, egalitarian, cosmopolitan community-world is envisioned. For the saints, the esoteric book that mandates formation of the new community is the Bible.

Wimbush's strongest indictment against the tradition was the description of their appropriation of the Bible as *bibliomancy,* which he defined as the reading of holy books for the purpose of solving personal problems or in order to effect some wonder from which one can benefit. The preaching, teaching, testimony, and songs of the Sanctified church tradition are a united witness of the efficacy of the Bible in the life of the believer. The sacred speech, song, and dance of the tradition are understood and performed with reference to specific texts. Not only do people experience salvation on the ground of biblical truth but also sanctification, Spirit baptism, healing, and deliverance are experiences mediated through the appropriation of Scripture. Thus, literal readings of Scripture are most fitting in view of the central significance ascribed to the Bible in these churches, notwithstanding the allegation that it is being treated as a book of magic. In a similar vein, Theophus Smith has offered a figural and realist alternative to literalist and Afrocentric appropriations of the Bible in his *Conjuring Cultures,* but his presentation of the Bible as a "conjurebook" is also fraught with difficulties.[34]

Leonard Barrett has commented upon the ethical and ecclesiological impact that biblical literacy produced in Africa. Once the Bible was translated and published in the vernacular languages, Africans began reading its vision of social renewal, power, prosperity, peace, love, justice, and equality. This new ethical awareness enabled them to discern a serious discrepancy between the Bible and the teaching and liturgical practices of the white missionaries. The next step was the emergence of the independent African Christian churches.[35] In view of this scenario, it seems feasible to suggest that the push for literacy and Bible reading among the freed slaves during the late nineteenth century may very well have been a major factor in the rise of the Sanctified church tradition as a biblically based alternative to the moral hypocrisy and liturgical boredom of white Christianity.

The gospel of Jesus Christ, as literally appropriated, interpreted, and promulgated by the saints, is the unifying message of the Sanctified church movement. To designate the movement's music as "gospel" music signifies not only its cultural style but also its ethical meaning and intent. From the vantage point of exile, it is apparent that the dialectics of reconciliation and refuge offer a vital platform for the promotion of the gospel—reconciliation as the rationale for reaching out to minister to persons across the barriers of race, class, and sex and refuge as the institutionalized objective of providing a place for those who have experienced marginalization and alienation to "come home."

The Sanctified church remains an important sphere of black identity. In many places within the tradition, the African particularity is not suppressed or concealed but rather is openly revealed as a witness to the universality of divine reality and creativity. The saints are African diasporic people giving liturgical expression to the pulse of the Spirit under conditions of exile, a people who are not ashamed to chant and cry and dance "Yes" to God. "Yes" is articulated not only in speech but also in the offbeat rhythms of Sanctified worship, in the clapped and danced antiphonal responses to the divine pulse "which beats the note that is never sounded."[36] The ethics of the Christian gospel sustains the churches' effort to love black people without despising whites.

Chancellor Williams's admonition that integration is needed *within* the black race, issued as an appeal for global black unity, was formulated out of his awareness of class discrimination in the denominational black churches, in contrast to the outreach of the storefront churches to the disinherited.[37] In confirmation of the current relevance of Williams's concern, Mary Sawyer concluded her recent monograph on the ecumenical activity of the black churches with the warning that "black ecumenism will empower and liberate only insofar as it includes within its purview the vast numbers of the black underclass and the marginal working class."[38]

It has been shown throughout this study that persons in the Holiness-Pentecostal movement have offered leadership in justice-oriented ecumenical efforts that focus on mission, advocacy, and intervention. Sometimes they have formed partnerships with those who may not otherwise endorse the movement's message. In early 1980s, for example, a citywide ecumenical Christian coalition spearheaded by Cheryl Sanders, pastor of First Church of God in Boston, Massachusetts; Eugene Rivers, student leader of the Seymour Society at Harvard University; and Francis Grady, a Roman

Catholic lay activist, mobilized residents of Roxbury and Dorchester to work with churches, college students, city political leaders, and law enforcement officials to drive drug dealers out of the Sonoma Street neighborhood. Rivers subsequently founded the Azusa Christian Community, which serves as an ecclesiastical base for youth outreach, community service, and political activism.[39]

Harvey Cox's *Fire from Heaven* provides an impressive survey of Pentecostalism as an urban ecumenical movement encompassing some 410 million adherents worldwide, or one of four Christians. Karla Poewe's anthology, *Charismatic Christianity as a Global Culture,* presents a host of insightful scholarly reflections upon significant currents and developments in global Pentecostalism. One contributor to that volume, Walter Hollenweger, has written a provocative essay on the global expansion of Pentecostalism, "The Pentecostal Elites and the Pentecostal Poor: A Missed Dialogue?" He saw the "oral quality" of Pentecostalism, rooted in the religious sensibilities of William J. Seymour, as the key to its phenomenal growth in the developing Third World:

> The oral quality of Pentecostalism consists of the following: orality of liturgy; narrative theology and witness; maximum participation at the levels of reflection, prayer, and decision making, and therefore a reconciliatory form of community; inclusion of dreams and visions into personal and public forms of worship that function as a kind of "oral icon" for the individual and the community; an understanding of the body-mind relationship that is informed by experiences of correspondence between body and mind as, for example, in liturgical dance and prayer for the sick.[40]

Hollenweger was highly critical of upwardly mobile and elite Pentecostals who subvert the ongoing global expansion of their religion among the poor by rejecting the very elements that made it possible for racial, social, and linguistic barriers to be overcome in the first place. He asked whether Pentecostals have the ability, the places, the language, the institutions, and the finances to present their theology lucidly and critically in oral categories just as Christ and the apostles did: "If we were able to do this, we would not only open up a dialogue on life and death with our own constituency among the poor, not to mention the religiously immune rich, we would also become a shining example for the Vatican and the World Council of Churches."[41] His concern for the alienation of the poor led to a critical questioning of the content of prophetic utterances heard in the Pentecostal services he attended:

> I have often asked myself why in our meetings the Spirit is so eloquent on "peace of heart," on marriage problems, and on questions of individual

ethics, and so silent on racism, oppression, and starvation? I do not think that the Spirit is unaware of these problems and that he does not want to lead us to better solutions than those of the politicians. Could it be that we do not listen or that we do not have the institutions and "places" where he, or she, can express him or herself? Could it be that by using a language that excludes the majority from our deliberations, we unconsciously have passed over "the least of these my brethren"? (Matthew 25:40).[42]

Thus, in the wake of nearly nine decades of explosive growth among the disinherited, some Pentecostals are beginning to repeat the mistakes of the mainstream churches: "We cut ourselves off from our own poor people in the interests of a streamlined theology, of an efficient organization and a facade of unity."[43]

In some key places outside the mainstream, however, there are vital manifestations of a "subversive and revolutionary movement," as Pentecostal scholar Cheryl Bridges Johns has observed, not based on philosophic ideology or totally on critical reflection but galvanized by the Holy Spirit's call for a radical equalizing of blacks and whites, males and females, the rich and the poor.[44] Empowered by an egalitarian, evangelistic synthesis of Christian ethics and African spirituality, the Sanctified churches present a robust ecumenical witness to exiles at home and abroad.

Conclusion: Exile and Homecoming

The saints in exile are religious communities of African Americans upon whom the North American Babylon has imposed alien status on account of their race, culture, class, and, in some cases, their sex. Moreover, they have further exiled themselves in significant ways by virtue of their codes of morality and their peculiar liturgies of song, speech, and dance. Exile has functioned in this study alternately as descriptive and normative concept. It is descriptive of the experience of African American people under the conditions of oppression and alienation. Its normative meaning is most fully revealed in the expressions of personal and social ethics that come to light as the exilic community worships.

Exile has been chosen here as a category of meaning for the interpretation of ethics and worship in African American religion and culture because it enables a more precise focus on intragroup identity and ethics than the exodus paradigm of black liberation theology affords by its analysis of victimization and suffering. Liberation theology emphasizes the moral obligation of the oppressor to set the oppressed free. An "exilic" theology would equally obligate the entire community of faith, inclusive of exiles and elites, to offer authentic liturgies of welcome and memory that enable the experience of liberation as homecoming.

In a variety of ways, twentieth-century African Americans have responded to the experience of exile and alienation in America by expressing their longing for some place or space—geographical, cultural, spiritual—where they can feel at home. African Americans constitute a unique category among North American exiles because they are the descendants of the only group of Old World people whose presence in the New World has been involuntary, as historian of religion Charles H. Long has observed, "having been brought to America in chains, and this country has attempted to keep them in this condition in one way or another."[1] Peter Paris has argued that because African Americans have been estranged from Africa and rejected by America, they were deprived of the two most basic existential conditions of life: a prideful place of origin and a firm sense of belonging. This existential insecurity fueled their hopeful pursuit of two alternative understandings of "home": a home in America in spite of their traditional exclusion, or a home in Africa based upon the reestablishment of roots after more than three centuries of cultural separation.[2] Notwithstanding these abiding hopes and aspirations, African Americans have been forced to live lives as perennial outsiders, finding a home only in a dynamic language and mobile music, as Cornel West has lamented, and never in a secure land, safe territory, or welcome nation. In this perspective, migration and emigration eclipse integration and separation as the fundamental themes of African American life.[3]

The plight of black intellectuals in this century has been to explicate the exile and alienation of their people in discourses of protest and to propose strategies toward the resolution of the crisis of black identity. In the latter part of the nineteenth century, Episcopal clergyman Alexander Crummell, physician Martin Delany, and African Methodist Episcopal Bishop Henry McNeal Turner spent time in Africa engaged in teaching, missions, or other activities and in exploration of the prospects for claiming an African homeland for African Americans.[4] Following his trip to the Niger Valley, Delany expressed "ambivalence over exile and homecoming," according to Paul Gilroy, a feeling associated historically with the African presence in the West: "Delany's African tour confirmed the dissimilarities between African-American ideologues and the Africans . . . though at the end of his account of his adventures in Africa Delany promised to return to Africa with his family, he never did so."[5] Marcus Garvey effectively communicated his vision of an African homecoming to the masses of black people in the 1920s. He prophesied the deliverance of his people from their exile in the New World and their restoration to Africa.[6] The agenda of Garvey's black nationalist organization, the Universal Negro Improvement Association, was to invite black people home, but he never succeeded in transporting himself or his followers to African soil.

Both W. E. B. Du Bois and E. Franklin Frazier died in the early 1960s, just as Old World and New World Africans were escalating their respective efforts to gain independence and civil rights. After a lifetime of thought and struggle, Du Bois and Frazier each envisioned some version of the African homecoming as a final solution to the experience of exile and alienation in the United States. Du Bois arranged a literal homecoming to Africa; he renounced his American citizenship and emigrated to Ghana in 1961 at the age of 93. He died there two years later, on August 27, 1963, the eve of the historic March on Washington. Frazier achieved his African homecoming only in the figurative sense, by means of a bequest of his library to the

University of Ghana, the home of Nkrumah's revolutionary nationalism and the site of what Frazier hoped might become "the birthplace of a new humanity."[7] On May 17, 1962, Frazier died in the United States after suffering a massive heart attack. Six weeks prior to his own death, Du Bois wrote Frazier's last epigraph in a message commemorating Frazier's bequest, declaring that "in the best sense of the words, E. Franklin Frazier was more fundamentally American than most Americans."[8] Yet, the United States State Department and the Federal Bureau of Investigation subjected both men to severe scrutiny and surveillance because of the subversive nature of their critical thought and activity.

As the twentieth century draws to a close, contemporary African American scholars are continuing this intellectual inquiry into the dual themes of exile and homecoming. Reflecting on his marriage to his Ethiopian wife, Elleni Gebre Amlak, Cornel West confessed his urge to leave America and live in exile in Ethiopia in the house provided by her mother.[9] His understanding of exile was set forth as a descriptive of the character of the prophetic criticism in which he is engaged—academically "unhoused," estranged from American society, marginal in relation to black culture, and "suspicious of any easy answers, quick fixes or dogmatic routes" to reach "home."[10]

Black feminist social theorist bell hooks described "homeplace" as a safe refuge from white domination and oppression, a site of resistance and liberation struggle, a place where "we can recover ourselves."[11] She has written about her own remembrances of childhood visits to the house she identifies as belonging to her grandmother (although her grandfather also lived there with her). In her youthful perspective, houses belonged to women, because the most important tasks performed there were the work of women: providing the comforts of shelter, food, and nurture of the soul; teaching dignity, integrity, and faith. Significantly, the context in which hooks located the highly esteemed task and tradition of "homeplace" in black women's culture is not Africa but a rural town in the southern United States.

This understanding of homeplace suggests that black men and women, including intellectuals of both sexes, have responded to the experience of exile in different ways. Moreover, as the writings of James Baldwin seem to signify, exile and homecoming may have special meaning for gays, lesbians, bisexuals, and other individuals who live apart from the conventional social structures of marriage and family.[12] The nature of the tasks hooks identified with homeplace—notably, teaching and character formation—signals the importance of the theological dimension of exile and homecoming, a factor sometimes ignored or devalued by black academicians pursuing secular approaches and private solutions to the problem of not belonging. Historian of religion Elias Farajajé-Jones has incorporated this theological dimension into his study of black religious movements and used exilic language to assert that blacks in North America have been constituted as a people in a process that engages their past, present, and future, "welded together by a past from which they had been uprooted, a present in which they were in captivity, and a future in which they were to be freed by the hand of God."[13]

The question of Psalm 137:4 — how could we sing the "Lord's song" in a "strange land"? — has been posed in this study to interrogate the exilic experience of African American Christians from the vantage point of the exilic narrative of the biblical Hebrews. Biblical scholar Walter Brueggemann's understanding of Babylonian exile and Israel's homecoming in light of the prophetic poetry of Jeremiah, Ezekiel, and second Isaiah is instructive for the effort to be guided by the Scriptures in bringing closure to this analysis of black exilic existence. In his book *Hopeful Imagination: Prophetic Voices in Exile*, Brueggemann described exile as a sense of not belonging, of being in an environment hostile to the values of this community and its vocation. It is practiced among those who refuse to accept and be assimilated in the new situation. Babylon is the empire that actively seeks to domesticate and oppress the exile, the concentration of power and value that is dominant and hostile to the covenant faith of the exiled community. Homecoming is a dramatic decision to break with imperial rationality and to embrace home as a place where covenantal values have currency and credibility.[14] With these definitions in hand, it is possible to show how the text raises questions that interrogate the exilic status of African American Christians.

"By the rivers of Babylon—there we sat down and there we wept when we remembered Zion . . . For there our captors asked us for songs" (Psalm 137:1, 3a, NRSV). Over against the Babylonians' demand to be serenaded by the exiles for entertainment, it is asserted that the only legitimate liturgical performances are those that bring forth memories of home to allow cathartic emotional release. That such power is inherent in ancestral black sacred music has been demonstrated by the Reverend William Herbert Brewster's recollection of the lyrics and emotion of a spiritual sung to him by his grandmother, a former slave, when Brewster was only three years of age. He can remember the words ("I am a poor pilgrim of sorrow . . . tell Hezekiah there is a lifetime journey everywhere we go") and the feelings: "She was singing and laughing out loud almost hysterically—that 'Holy Ghost laugh,' they used to call it. And then she would be crying. The tears were meeting under her chin. It touched my heart so deeply."[15] Brewster grew up to become a Baptist pastor and gospel songwriter in Memphis, the city of W. C. Handy, Mason Temple (mother church of the Church of God in Christ), and Elvis Presley, who frequently visited Brewster's congregation to hear the music. Anthony Heilbut described Brewster's songs as "a seamless blend of Baptist and Methodist decorum and Sanctified ecstasy."[16]

Brewster is credited with writing the first two gospel songs to result in the sale of more than a million records: Mahalia Jackson's signature performance, "Move On Up a Little Higher" (1947) and "Surely God Is Able" (1950) by Clara Ward and the Ward Singers, with Marion Williams singing the lead. Both songs seem to have appealed to the masses of black people as anthems of struggle and pilgrimage. In "Surely God Is Able," the recitation of numerous biblical stories of divine deliverance is an affirmation of God's power, God's care and concern for the desolate pilgrim, God's presence in difficult circumstances, and God's overall sustaining power. Cheryl Townsend Gilkes viewed this song as an Afrocentric folk text that points to the vitality of the Afro-Christian biblical tradition. Verifying its importance in the evolution of a folk tradition in Afrocentric biblical hermeneutics, she noted how the song employs several hermeneutical voices as it combines a "basic affinity for the liberating dimensions of the Hebrew Bible with an insistence that images of God be affirmed in the experience of black persons and their community."[17]

"How could we sing the LORD's song in a foreign land? If I forget you, O Jerusalem. . . . Let my tongue cling to the roof of my mouth." (Psalm 137:4–6a, NRSV). Brueggemann discerned in Psalm 137 a "passionate resolve not to be assimilated."[18] At the root of the question raised in verse 4 is a theological pretext and imperative for the exile to resist assimilation of the oppressor's will and ways. Here resistance becomes a matter of concern that is expressly theological. In his published sermon "Faith in a Foreign Land," pastor-scholar Jeremiah Wright of the Trinity United Church of Christ in Chicago drew freely on the biblical exilic texts to expose the absurdity of black assimilation of white values: "You will have African exiles who think that unless the Babylonians said it, it ain't true; unless Babylonians wrote it, it ain't right; unless the Babylonians made it, it ain't gonna work."[19] Regarding the necessity of "remembering Zion," Wright asserted that in a foreign land an identity crisis is created in a deliberate attempt to take away the exiles' history and then to completely destroy their heritage.

"O daughter Babylon, you devastator! Happy shall they be who pay you back what you have done to us! Happy shall they be who take your little ones and dash them against the rock! (Psalm 137:8–9, NRSV). The last verses of Psalm 137 take the form of a prayer for God to punish and destroy Babylon for crimes committed against the people of God. This "curse prayer" is offered in a spirit of sadistic vengeance and invokes a theological strategy for employing violence in response to feelings and memories of despair.

Generally speaking, the African American prayer tradition has been tempered in this regard by the Christian ethic of eschewing violence and hatred toward one's enemies. Even David Walker's fiery antebellum exhortations to slave rebellion were motivated by a resolve to honor the demands of divine justice and mercy in formulating morally appropriate responses to the global phenomenon of white racist imperialism. Notwithstanding his confession of hatred of his people's white enemies, his messianic vision of a "united and happy America" calls for racial reconciliation at home and rejects colonization in Africa.[20] Walker's contemporary, Maria Steward, drew upon the imagery of the exodus in presenting her case against white America before God:

> You may kill, tyrannize, and oppress as much as you choose, until our cry shall come up before the throne of God; for I am firmly persuaded, that he will not suffer you to quell the proud, fearless and undaunted spirits of the Africans forever; for in his own time, he is able to plead our cause against you, and to pour out upon you the ten plagues of Egypt.[21]

The concern Walker and Steward showed for the preservation of human dignity and the promotion of ideals of racial equality in response to violence directed toward blacks would be reiterated in this century by Howard Thurman and by Martin Luther King Jr., who promoted a Christian philosophy of nonviolence as a strategic response to the perennial problem of racial injustice that frames African American exilic existence. The analysis of black despair, however, has not gone unheeded by black intellectuals confronted with the nihilism of the black urban underclass and the identity crisis of the black middle class.[22]

Brueggemann's interpretation of the exilic literature of the Bible speaks specifically to the social and moral significance of class differences in modern African American life. He assessed the efforts undertaken by the exilic prophets and poets to convince the Jews to remember from whence they had come and to resist assimilation of the values and gods of Babylon; he concluded that the combination of taking Babylonian definitions of reality too seriously and the loss of the peculiar memories of the faith community leads to life-destroying despair.[23] Assimilation and amnesia are equally problematic for the rich and the poor of modern-day America. This formula for despair plagues the rich and poor of all races, prompting the former toward increased repression and denial of the deleterious effects of racism (as exemplified by the rhetoric of contemporary black conservative leaders) and producing destructive self-hatred and violence among the latter. The only effective remedy for exilic despair is the liturgical rehearsal of sacred truths concerning the preexilic past and the promised homecoming of the future. Such "future-giving memory" is the main thing an established empire has to fear.[24]

Two important African American sources of future-giving memory were suggested by Brueggemann: the slave narrative and gospel music. He quoted Alice Walker's opinion concerning the importance of the slave narratives in causing struggle and freedom to emerge as distinctive themes in black literature: "Black writers seem always involved in a moral and/or physical struggle, the result of which is expected to be some larger kind of freedom. Perhaps this is because our literary tradition is based on the slaves' narratives, where escape for the body and freedom for the soul went together."[25] The slave experience is communicated as future-giving memory primarily in the spirituals and to some extent in the folk tradition of black preaching.

William Herbert Brewster's account of how and why he wrote "Move On Up a Little Higher" provides an illustration of the exilic symbolism and subversive nature of the gospel song as "future-giving memory":

> Before the freedom fights started, before the Martin Luther King days, I had to lead a lot of protest meetings. In order to get my message over, there were things that were almost dangerous to say, but you could sing it.
>
> "Move on up a Little Higher" was not only a Christian climbing the ladder to heaven, but it was an exaltation of Black people to keep moving. You come out of slavery, you have an opportunity to get on your ladder and keep on climbing. Don't stop when you make one; make another. That was a general idea that became popular because that had a lot of bounce in it. Mahalia [Jackson] knew what to do with it.[26]

What Mahalia Jackson did with it was to make a recording in 1947 that sold more than a million copies, enabling her to come into her own as the Queen of Gospel. The songwriters Dorsey and Brewster, together with Jackson in the role of "caller," can be seen as exilic poets of the "urban bush arbor," who created sung liturgies with mass appeal to summon their people home within the American context of racial alienation.[27] As African exiles "came to themselves" in twentieth-century America, many heeded the call of Dorsey and the prophetic poets of the Sanctified church to find sanctuary in true worship.

The verse in Jesus' parable of the two lost sons, in which the prodigal son "came to himself" (Luke 15:17), was cited by Brueggemann as a subtext for a fresh proclamation of the gospel to the exiled church in America. In his view, the "gospel" is that we may go to a home not situated in the "consumer militarism" of the dominant value system but rather in God's kingdom of love and justice and peace and freedom. Brueggemann declared: "The news is we are invited home. The whole church may yet sing: 'Precious Saviour take my hand. Lead me home!' "[28] His allusion, of course, was to Thomas A. Dorsey's "Take My Hand, Precious Lord," the beloved gospel song that has become for African Americans an anthem of exilic existence and aspiration, often sung at funerals. It is a prayer to be divinely led through a dark and treacherous journey to a place called home. Notwithstanding his invocation of the Scriptures and this song, somehow Brueggemann did not seem confident that white American Christians will ever join black American Christians in embracing the identity of exile as a "whole church."

The appeal of the exilic metaphor to white Christians remains problematic. At the end of *Fire from Heaven*, Harvey Cox claimed to have "come full circle" in his examination of global Pentecostalism when he visited the Azusa Christian Community in Dorchester, Massachusetts, located just a few miles from his own home in Cambridge.[29] After having traveled virtually around the globe seeking understanding of Pentecostalism as a worldwide religious movement, it is somewhat ironic that he discovered his most satisfying validation of the "original spirit of Azusa Street" so close to home. Apparently there is no true "homecoming" for the white American prodigal in the fellowship of his black Pentecostal neighbors, however, only a warm welcome and an informative visit.

As a faith community whose dual heritage is rooted in the evangelical imperatives of holiness and unity associated with the nineteenth-century egalitarian Holiness movement and in the fires of Azusa Street that set the Pentecostal revival ablaze in the twentieth century as a global, multicultural phenomenon, the Sanctified church in America is challenged to sustain a socially conscious and profoundly spiritual Christian witness.[30] The worship and music of the Sanctified churches embody a host of ethical responses to the exilic existence imposed on African Americans as a consequence of white racism, both in the church and outside it. The genius of this embodied ethics is that it promotes racial reconciliation without obliterating racial identity. Moreover, the witness of those saints may have helped to save African American folk culture from total oblivion in this century because they stood against black middle-class assimilationism without apology and because they saw the blues not as a threat to black piety but as a cathartic vehicle for giving theodicy full liturgical voice.

The future of the tradition seems imperiled, however, by the drift toward the acceptance of elitist patriarchal values, the rejection of ancestral identity and worship practices, and the failure to sustain a creative evangelistic outreach in the face of the host of nihilistic encroachments of despair—homelessness, poverty, drug abuse, crime, family dysfunction—that have become endemic to black urban life. The task at hand is to find more effective ways to communicate the gospel of Jesus Christ to those most at risk to lose their way in the current landscape.

At places like the Third Street Church of God in Washington, D.C., the homeless are invited to "have church" with the privileged in worship experiences that encourage repentance and totalistic transformation. Perhaps the best hope the Sanctified church tradition has to offer to the world is the preservation of a divinely anointed and authorized space in this North American Babylon and beyond, where the privileged can deal with their resentments and the dispossessed can dispel their doubts and where both can learn to celebrate homecoming with compassion and conviction. Under the influence of the sacred practices of the slave ancestors, may the ethics and worship of the Sanctified church continue to connect the dark memory of exilic suffering with the bright promise of homecoming. Yes, Lord!

Notes

Introduction

1. Zora Neale Hurston, *The Sanctified Church* (Berkeley: Turtle Island, 1981), 103.

2. Ibid., 104–107.

3. Cheryl Townsend Gilkes, "The Role of Women in the Sanctified Church," *Journal of Religious Thought* 43, no. 1 (spring-summer 1986), 25, and Gilkes, "Together and in Harness: Women's Traditions in the Sanctified Church," *Signs: Journal of Women in Culture and Society* 10, no. 4 (summer 1985), 679–680.

4. Wardell Payne, ed., *Directory of African American Religious Bodies* (Washington, D.C.: Howard University Press, 1991), 82–116.

5. William C. Turner Jr., "Movements in the Spirit: A Review of African American Holiness/Pentecostal/Apostolics," in Payne, *Directory*, 248.

6. Ibid.

7. Leonard Lovett, "Black Holiness-Pentecostalism: Implications for Ethics and Social Transformation" (Ph.D. dissertation, Emory University, 1978), 13.

8. Turner, "Movements in the Spirit," 253.

9. Molefi Kete Asante, *Afrocentricity* (Trenton, N. J.: Africa World Press, 1988), 74.

10. Ibid.

11. Ibid.

12. James S. Tinney, "A Theoretical and Historical Comparison of Black Political and Religious Movements" (Ph.D. dissertation, Howard University, 1978), 232–233.

13. Ibid., 234.

14. Ibid., 236–237.

15. James S. Tinney, "The Blackness of Pentecostalism," *Spirit* 3, no. 2 (1979), 29.

16. John Edward Philips, "The African Heritage of White America," in *Africanisms in American Culture*, ed. Joseph E. Holloway (Bloomington: Indiana University Press, 1990), 231.

17. Ibid., 237.

18. Ibid., 227.

19. Joseph M. Murphy, *Working the Spirit: Ceremonies of the African Diaspora* (Boston: Beacon Press, 1994), 4.

20. Ibid., 146.

21. Ibid.

22. Ibid., 2.

23. Ibid., 2–3.

24. See Dwight Hopkins and George Cummings, eds., *Cut Loose Your Stammering Tongue: Black Theology in the Slave Narratives* (Maryknoll, N.Y.: Orbis, 1990); Thomas L. Webber, *Deep Like the Rivers: Education in the Slave Quarter Community, 1831–1865* (New York: W. W. Norton, 1978); Clifton Johnson, ed., *God Struck Me Dead* (1969; reprint, Cleveland: Pilgrim Press, 1993); Charles L. Perdue Jr., et. al., eds., *Weevils in the Wheat* (Bloomington: Indiana University Press, 1976); and Cheryl J. Sanders, "Slavery and Conversion: An Analysis of Ex-Slave Testimony" (Th.D. dissertation, Harvard University, 1985).

25. Sarah Bradford, *Harriet Tubman: The Moses of Her People* (New York: Corinth Books, 1961), 23, 24.

26. Ibid., 25.

27. Ibid., 26.

28. Ibid., 29.

29. Ibid., 31–32.

30. Albert J. Raboteau, *Slave Religion* (New York: Oxford University Press, 1978), 72.

31. Daniel Alexander Payne, *Recollections of Seventy Years* quoted in Ibid., 68. 1st publ. 1886 (New York: Arno Press and the New York Times, 1969), 253–55.

32. Raboteau, *Slave Religion,* 86.

33. Ibid., 74.

34. Ibid., 314.

35. Ibid., 301.

36. Ibid., 302.

37. Ibid., 224.

38. Ibid., 307–309.

39. Sterling Stuckey, *Slave Culture: Nationalist Theory and the Foundations of Black America* (New York: Oxford University Press, 1987), 95.

40. Ibid., ix.

41. Bradford, *Harriet Tubman,* 92–93.

42. David Douglas Daniels III, "The Cultural Renewal of Slave Religion: Charles Price Jones and the Emergence of the Holiness Movement in Mississippi" (Ph.D. dissertation, Union Theological Seminary, 1992), 89.

43. Ibid., 16.

44. Ibid., 234.

45. Ibid., 185.

46. Ibid., 268.

Chapter 1

1. Leonard Lovett, "Black Holiness-Pentecostalism: Implications for Ethics and Social Transformation (Ph.D. dissertation, Emory University, 1978), 13.

2. William C. Turner Jr., "Movements in the Spirit: A Review of African American Holiness/Pentecostal/Apostolics," in *Directory of African American Religious Bodies*, ed. Wardell Payne (Washington, D.C.: Howard University Press, 1991), 250.

3. Cornel West, "The Prophetic Tradition in Afro-America," in *Prophetic Fragments* (Grand Rapids, Mich.: Wm. B. Eerdmans, 1988), 43.

4. Turner, "Movements in the Spirit," 249.

5. Albert J. Raboteau, *Slave Religion* (New York: Oxford University Press, 1978), 139.

6. Turner, "Movement in the Spirit," 249. See Harry V. Richardson, *Dark Salvation: The Story of Methodism As It Developed among Blacks in America* (New York: Anchor/Doubleday, 1976).

7. C. Eric Lincoln and Lawrence H. Mamiya, *The Black Church in the African American Experience* (Durham, N. C.: Duke University Press, 1990), 50–51.

8. Ithiel C. Clemmons, "Charles Mason Harrison," in *Dictionary of Pentecostal and Charismatic Movements*, ed. Stanley M. Burgess and Gary B. McGee (Grand Rapids, Mich.: Zondervan, 1988), 587.

9. James S. Tinney, "The Blackness of Pentecostalism," *Spirit* 3, no. 2 (1979), 30.

10. Information on each of these church bodies taken from entries appearing in Burgess and McGee, eds., *Dictionary of Pentecostal and Charismatic Movements* and Wardell Payne, ed., *Directory of African American Religious Bodies*.

11. Payne, *Directory*, 206–208.

12. Statistics cited in *1995 Yearbook of the Church of God* (Anderson, Ind.: Leadership Council of the Church of God, 1995), 337, and *Commemorative Booklet in Observance of the Centennial Celebration of the Church of God, 1880–1980* (West Middlesex, Pa.: National Association of the Church of God, 1981), 87.

13. Katie H. Davis, ed., *Zion's Hill at West Middlesex* (1951; reprint, Anderson, Ind.: Shinning Light Press, 1985), 10–11.

14. Ibid., 27.

15. Ibid., 28.

16. Ibid., 11.

17. Ibid., 18.

18. Ibid., 21–22.

19. Ibid., 23.

20. Ibid., 46.

21. Ibid., 24.

22. Ibid., 19.

23. Cheryl Townsend Gilkes, " 'Together and in Harness': Women's Traditions in the Sanctified Church," *Signs* 10, no. 4 (Summer, 1985), 695.

24. Davis, *Zion's Hill*, 12.

25. Ibid., 29–30.

26. Ibid., 46.

27. John W. V. Smith, *The Quest for Holiness and Unity* (Anderson, Ind.: Warner Press, 1980), 96.

28. Davis, *Zion's Hill*, 26.

29. Samuel G. Hines, with Joe Allison, *Experience the Power* (Anderson, Ind.: Warner Press, 1993), 91.

30. Davis, *Zion's Hill*, 47.

31. Ibid., 47.

32. William E. Montgomery, *Under Their Own Vine and Fig Tree: The African-American Church in the South, 1865–1900* (Baton Rouge: Louisiana State University Press, 1993), 349.

33. James S. Tinney, "A Theoretical and Historical Comparison of Black Political and Religious Movements" (Ph.D. dissertation, Howard University, 1978), 221.

34. H. Vinson Synan, "William Joseph Seymour," in Burgess and McGee, *Dictionary,* ed. 780.

35. Montgomery, *Under Their Own Vine*, 349.

36. Synan, "Seymour," 781.

37. James S. Tinney, "William J. Seymour: Father of Modern-Day Pentecostalism," in *Black Apostles,* ed. Randall K. Burkett and Richard Newman (Boston: G. K. Hall, 1978), 217.

38. This account of Seymour's role in the beginnings of the Azusa Street Revival is adapted from several sources. See Joseph Colletti, *Selected Historical Pentecostal Sites in the Los Angeles Area* (Pasadena, Calif.: David J. du Plessis Center for Christian Spirituality; undated handout); Leonard Lovett, "Aspects of the Spiritual Legacy of the Church of God in Christ: Ecumenical Implications" in *Black Witness to the Apostolic Faith,* ed. David T. Shannon and Gayraud S. Wilmore (Grand Rapids, Mich.: Wm. B. Eerdmans, 1988); Cecil M. Robeck, Jr., "Azusa Street Revival" and "Bonnie Brae Street Cottage" in *Dictionary,* ed. Burgess and McGee; Synan, "Seymour"; and Tinney, "Seymour."

39. "Weird Babel of Tongues," *Los Angeles Daily Times,* April 18, 1906.

40. Turner, "Movements in the Spirit," 251.

41. Ibid.

42. James R. Goff Jr., "Charles Fox Parham," in *Dictionary,* ed. Burgess and McGee, 661.

43. Leonard Lovett, "Perspective on the Black Origins of the Contemporary Pentecostal Movement," *Journal of the I.T.C.* 1, no. 1 (1973), 46.

44. Synan, "Seymour," 781.

45. Ibid.

46. Robeck, "Azusa Street Revival," 35.

47. Colletti, "Selected Historical Pentecostal Sites," n.p.

48. Tinney, "Seymour," 222.

49. Memphis *Commercial Appeal,* May 22, 1907, quoted in David M. Tucker, *Black Pastors and Leaders: Memphis 1819–1972* (Memphis: Memphis State University Press, 1975), 90–91.

50. Ibid., 97.

51. Roger Finke and Rodney Stark, *The Churching of America, 1776–1990: Winners and Losers in Our Religious Economy* (New Brunswick, N. J.: Rutgers University Press, 1992), 237.

52. For point-by-point comparison with the Apostolic Faith statement, see F. G. Smith, *What the Bible Teaches,* condensed by Kenneth E. Jones (1914; reprint, Anderson, Ind.: Warner Press, 1955), chapter 6, "Conditions for Salvation" (especially subject headings Godly Sorrow, Repentance, Confession, Restitution, and Forgiveness), 36–41; chapter 9, "Sanctification as a Bible Doctrine," 63–74; chapter 10, "Divine Healing," 77–84; and chapter 11, "The Unity of Believers," 85–98. The restoration of the church is discussed in chapter 19, "The True Church Restored," 129–138.

53. Turner, "Movements in the Spirit," 251.

54. William J. Seymour, quoted in Ian MacRobert, *The Black Roots and White Racism of Early Pentecostalism in the USA* (London: Macmillan Press, 1988), 48.

55. John W. V. Smith, *A Brief History of the Church of God Reformation Movement* (Anderson, Ind.: Warner Press, 1976), 122.

56. Pearl Williams-Jones, "A Minority Report: Black Pentecostal Women," *Spirit* 1, no. 2 (1977), 31–44.

57. The plight of women ministers in the Church of God is addressed by several authors in Juanita Leonard, ed., *Called to Minister, Empowered to Serve* (Anderson, Ind.: Warner Press, 1989).

58. Susie Stanley, quoted by Timothy C. Morgan, "The Stained-Glass Ceiling," *Christianity Today* 38, no. 6, (May 16, 1994), 52.

59. Susie Cunningham Stanley, *Feminist Pillar of Fire: The Life of Alma White* (Cleveland; Pilgrim Press, 1993), 2.

60. Leonard Lovett, "Aspects of the Spiritual Legacy of the Church of God in Christ: Ecumenical Implications" in *Black Witness,* ed. Shannon and Wilmore, 47.

61. James S. Tinney, "The Prosperity Doctrine: Perverted Economics," *Spirit* 2, no. 1 (1978), 45.

Chapter 2

1. The detailed history of this congregation appears in Cheryl J. Sanders, *How Firm a Foundation: Eighty Years of History at Third Street Church of God (1910–1990)* (Washington, D.C.: Third Street Church of God, 1990). The account of the congregation's origins was given by Lucille Young Baguidy, who is granddaughter and niece of two of the female founders of the congregation, on the occasion of the seventy-fifth anniversary of the Third Street Church of God, December 14, 1985, quoted in Sanders, *How Firm a Foundation,* 1.

2. Lewis Reed, in Sanders, *How Firm a Foundation*, 10.

3. Samuel G. Hines, *How Firm a Foundation*, 32.

4. *1994 Yearbook of the Church of God* (Anderson, Ind.: Leadership Council of the Church of God, 1994), 61.

Chapter 3

1. The author's observations concerning worship are based in part on data gathered from 1990 to 1994 during visits to seventy-five churches and twenty-eight college campus worship settings in twenty-one states (and the District of Columbia) in every region of the United States, representing twenty-five mainline Protestant, Catholic, Pentecostal, and Holiness denominations.

2. James Maynard Shopshire, "A Socio-historical Characterization of the Black Pentecostal Movement in America" (Ph.D. dissertation, Northwestern University, 1975), 170–183.

3. Arthur E. Paris, *Black Pentecostalism: Southern Religion in an Urban World* (Amherst: University of Massachusetts Press, 1982), 54–70.

4. Joseph M. Murphy, *Working the Spirit: Ceremonies of the African Diaspora* (Boston: Beacon Press, 1994), 158–169.

5. Wynton Marsalis, quoted by Stanley Crouch, "In the Sweet Embrace of Life," liner notes to compact disk recording by Wynton Marsalis Septet, *In This House, on This Morning* (Columbia C2K53220 1994), 5–6.

6. Melva Wilson Costen, *African American Christian Worship* (Nashville: Abingdon Press, 1993), 136–140. See also her brief discussion of Pentecostal and Holiness worship on pp. 113–115.

7. Shopshire, "Socio-historical Characterization," 172.

8. Bernice Johnson Reagon, "Searching for Tindley," in *We'll Understand It Better By and By: Pioneering African American Gospel Composers* (Washington, D.C.: Smithsonian Institution Press, 1992), 39–40.

9. Interview with Kenneth Morris, "I'll Be a Servant for the Lord," conducted and edited by Bernice Johnson Reagon, in *We'll Understand It,* 338.

10. Harold Dean Trulear, "The Lord Will Make a Way Somehow: Black Worship and the Afro-American Story," *Journal of the Interdenominational Theological Center* 12, no. 1 (fall 1985), 100.

11. James S. Tinney, "A Theoretical and Historical Comparison of Black Political and Religious Movements" (Ph.D. dissertation, Howard University, 1978), 240–241.

12. One is encouraged in the pursuit of this line of thinking in view of Rudolf Otto's classic phenomenological study of religion, first published in English in 1923, *The Idea of the Holy* (reprint, trans. John W. Harvey, New York: Oxford University Press, 1978). In an effort to define and describe the concept of holiness in worship, Otto introduces the term *mysterium tremendum,* which he understands as having two distinct modes of manifestation:

Let us consider the deepest and most fundamental element in all strong and sincerely felt religious emotion. Faith unto salvation, trust, love—all these are there. But over and above these is an element which may also on occasion, quite apart from them, profoundly affect us and occupy the mind with a well nigh bewildering strength . . . in sudden strong ebullitions of personal piety and the frames of mind such ebullitions evince, in the fixed and ordered solemnities of rites and liturgies. . . . It may burst in sudden eruption up from the depths of the soul with spasms and convulsions, or lead to the strangest excitements, to intoxicated frenzy, to transport, and to ecstasy. (12–13)

13. Trulear, "The Lord Will Make a Way," 96–97.

14. Katrina Hazzard-Gordon, "Dancing to Rebalance the Universe: African American Secular Dance and Spirituality," *Black Sacred Music* 7, no. 1 (spring 1993), 17.

15. Ibid., 19.

16. Ibid., 19.

17. Shopshire, "Socio-historical Characterization," 180–181.

18. Winthrop S. Hudson, "Shouting Methodists," *Encounter* 29, no. 1 (winter 1968), 84.

19. Ibid., 82.

20. Ann Taves, "Knowing through the Body: Dissociative Religious Experience in the African- and British-American Methodist Traditions," *Journal of Religion* 73, no. 2 (April 1993), 209.

21. Ibid., 215.

22. Ibid., 213.

23. Ibid., 218.

24. Pearl Williams-Jones, "The Musical Quality of Religious Folk Ritual," *Spirit* 1, no. 1 (1977), 29.

25. Murphy, *Working the Spirit,* 160.

26. My grandfather Brother Ellis Haizlip and several other men at Third Street Church of God served regularly during the 1970s as members of St. Luke's Guild. It was established in 1972 by my great-aunt Sister Levolia Goolsby, a licensed practical nurse, for the purpose of rendering first aid and other assistance as needed during the worship services. The original male and female members of St. Luke's Guild received first aid training by the local chapter of the Red Cross, and they wore white nurse's and orderly's uniforms while serving during worship. The persons most active in St. Luke's Guild at present are graduate registered nurses. One of the nurses, who is also a minister, wears white at every service, prays for and visits the sick, and participates in the rituals of anointing with oil and laying on of hands.

27. Nannie Helen Burroughs, "Who Started Women's Day?" in *Women and Religion in America*, 3, 1900–1968, ed. Rosemary Radford Ruether and Rosemary Skinner Keller (San Francisco: Harper & Row, 1986), 121.

28. C. Eric Lincoln, *The Black Church since Frazier* (New York: Schocken Books, 1974), 116.

Chapter 4

1. C. Eric Lincoln and Lawrence H. Mamiya, *The Black Church in the African American Experience* (Durham, N. C.: Duke University Press, 1990), 378–379.

2. Ibid., 381.

3. Ibid., 377.

4. Portia K. Maultsby, "The Impact of Gospel Music on the Secular Music Industry," in *We'll Understand It Better By and By,* ed. Bernice Johnson Reagon (Washington, D. C.: Smithsonian Institution Press, 1992), 24.

5. Thomas A. Dorsey, "Gospel Songwriter Attacks All Hot Bands' Swinging Spirituals," *Chicago Defender* (February 8, 1941), 2; reprinted in *Black Sacred Music* 7, no. 1 (spring 1993), 29.

6. Maultsby, "Impact of Gospel Music," 26.

7. For an interesting analysis of the negative portrayals of diasporic religions in film see Joseph M. Murphy's "Black Religion and 'Black Magic': Prejudice and Projection in Images of African-Derived Religions," *Religion* 20 (1990), 323–337.

8. Among the most useful sources on the history of gospel music are Michael W. Harris, *The Rise of Gospel Blues: The Music of Thomas Andrew Dorsey in the Urban Church* (New York: Oxford University Press, 1992); Anthony Heilbut, *The Gospel Sound: Good News and Bad Times* (1971, rev. ed. 1985); Irene V. Jackson, *Afro-American Religious Music: A Bibliography and Catalogue of Gospel Music* (1979); Bernice Johnson Reagon, ed., *We'll Understand It;* Eileen Southern, *The Music of Black Americans* (1971, 2d ed. 1982); Jon Michael Spencer, *Protest and Praise: Sacred Music of Black Religion* (1990); and Wyatt Tee Walker, *Somebody's Calling My Name: Black Sacred Music and Social Change* (1979).

9. Pearl Williams-Jones, "Afro-American Gospel Music: A Crystallization of the Black Aesthetic," *Ethnomusicology* 19, no. 3 (September 1975), 376.

10. Harris, *Rise of Gospel Blues,* 151.

11. Kenneth W. Osbeck, *101 Hymn Stories* (Grand Rapids, Mich.: Kregel Publications, 1982), 31.

12. Heilbut, *Gospel Sound,* 332.

13. Ibid., 248.

14. Albert J. Raboteau, *Slave Religion* (New York: Oxford University Press, 1978), 243.

15. Ibid., 243.

16. Harris, *Rise of Gospel Blues,* 69.

17. Ibid., 248.

18. Ibid., 247.

19. Williams-Jones, "Afro-American Gospel Music," 376.

20. Thomas A. Dorsey, quoted in Harris, *Rise of Gospel Blues,* 60.

21. Ibid., 239.

22. Thomas A. Dorsey, quoted in ibid., 241.

23. Adapted from the definition of *jazz* in *The American Heritage Dictionary of the English Language*, ed. William Morris (New York: American Heritage Publishing Co., Inc., 1975), 702.

24. Leonard E. Barrett, *Soul Force* (Garden City, N. Y.: Doubleday, 1974), 82.

25. Ibid., 83.

26. Sterling Stuckey, *Slave Culture: Nationalist Theory and the Foundations of Black America* (New York: Oxford University Press, 1987), 60.

27. Ibid., 95.

28. Harvey Cox, "Jazz and Pentecostalism," *Archives de Sciences Sociales des Religions* 84 (Octobre-Decembre 1993), 181–182. See also his *Fire from Heaven* (Reading, Mass.: Addison-Wesley, 1995), 143–144.

29. Portia K. Maultsby, "Africanisms in African-American Music," in *Africanisms in American Culture,* ed. Joseph E. Holloway (Bloomington: Indiana University Press, 1990), 201.

30. Heilbut, *Gospel Sound,* 175–176.

31. Cheryl Townsend Gilkes, "The Role of Women in the Sanctified Church," *Journal of Religious Thought* 43, no. 1 (spring-summer, 1986), 25.

32. Cornel West, "On Afro-American Popular Music: From Bebop to Rap," in *Prophetic Fragments* (Grand Rapids, Mich.: Wm. B. Eerdmans, 1988), 177–178.

33. Stuckey, *Slave Culture,* 96.

34. Cox, "Jazz and Pentecostalism," 187.

35. Wynton Marsalis, quoted by Stanley Crouch, "In the Sweet Embrace of Life," liner notes to compact disk recording by Wynton Marsalis Septet, *In This House, on This Morning* (Columbia C2K53220, 1994) 4.

36. Ibid., 8.

37. Ibid., 2.

38. Reuben Jackson, "Marsalis's Spirited 'House,' " *Washington Post,* June 22, 1994.

39. Geoffrey Himes, " 'R&B Box': Slightly out of Rhythm," *Washington Post,* December 14, 1994.

40. Morton Marks, "Uncovering Ritual Structures in Afro-American Music," in *Religious Movements in Contemporary America,* eds. Irving Zaretsky and Mark P. Leone (Princeton, N. J.: Princeton University Press, 1974), 114.

41. Reebee Garofalo, "Crossing Over: 1939–1989," in *Split Image: African Americans in the Mass Media,* ed. Jannette L. Dates and William Barlow (Washington, D.C.: Howard University Press, 1990), 69.

42. Marks, "Uncovering Ritual Structures," 114.

43. West, "On Afro-American Popular Music," 178–179.

44. Cornel West, "In Memory of Marvin Gaye," in *Prophetic Fragments,* 175.

45. Geoffrey Himes, "Return of the Gospel Truth," *Washington Post,* June 26, 1994.

46. Pearl Williams-Jones, "The Musical Quality of Religious Folk Ritual," *Spirit* 1, no. 1 (1977), 29.

47. Pearl Williams-Jones, "Afro-American Gospel Music," 381.

48. Morton Marks, " 'You Can't Sing Unless You're Saved': Reliving the Call in Gospel Music," in *African Religious Groups and Beliefs: Papers in Honor of William R. Bascom,* ed. Simon Ottenberg (Meerut, India: Archane Publications, 1982), 315.

49. James Weldon Johnson, *God's Trombones: Seven Negro Sermons in Verse* (1927; reprint, New York: Penguin Books, 1990), 11.

50. West, "On Afro-American Popular Music," 185.

51. Ibid., 185–186.

52. Ibid., 187.

53. Michel Marriott, "Rhymes of Redemption," *Newsweek,* November 28, 1994, 64.

54. Both Pearl Williams-Jones and Richard Smallwood were trained at the Howard University College of Fine Arts.

55. See Jon Michael Spencer, *As the Black School Sings,* Music Reference Collection no. 13 (Westport, Conn.: Greenwood Press, 1987).

56. Sallie Martin, quoted in Heilbut, *Gospel Sound,* 6.

57. Horace Clarence Boyer, "Kenneth Morris, Composer and Dean of Black Gospel Music Publishers," in *We'll Understand It,* ed. Reagon, 311–312.

58. Harris, *Rise of Gospel Blues,* 256.

59. Ibid., 262.

60. Heilbut, *Gospel Sound,* 17–18.

61. Ibid., 15–16.

62. Ibid., 196.

63. Ibid., 199.

64. Ibid., 200.

65. Brian Lanker, *I Dream a World* (New York: Stewart, Tabori & Chang, 1989), 78.

66. Heilbut, *Gospel Sound,* 202.

67. Lanker, *I Dream a World,* 78.

68. Violinist Marilyn E. Thornton Tribble, whose husband, Sherman Tribble, is a Baptist pastor, shared this insight with the author in a conversation concerning the secular nature of sacred music.

69. Heilbut, *Gospel Sound,* 200.

Chapter 5

1. David Douglas Daniels III, "The Cultural Renewal of Slave Religion: Charles Price Jones and the Emergence of the Holiness Movement in Mississippi" (Ph.D. dissertation, Union Theological Seminary, 1992), 276.

2. Cheryl Townsend Gilkes, " 'Together and in Harness': Women's Traditions in the Sanctified Church," *Signs* 10, no. 4 (summer 1985), 693.

3. Ibid., 697.

4. Walter J. Hollenweger, "Pentecostalism and Black Power," *Theology To-day* 30, no. 3 (October 1973), 230.

5. Luther P. Gerlach and Virginia H. Hine, *People, Power, Change: Movements of Social Transformation* (Indianapolis: Bobbs-Merrill, 1970), 204.

6. James S. Tinney, "A Theoretical and Historical Comparison of Black Political and Religious Movements" (Ph.D. dissertation, Howard University, 1978), 289.

7. C. Eric Lincoln and Lawrence H. Mamiya, *The Black Church in the African American Experience* (Durham, N. C.: Duke University Press, 1990), 363.

8. Alton B. Pollard III, *Mysticism and Social Change: The Social Witness of Howard Thurman* (New York: Peter Lang, 1992), 3.

9. Stephen N. Short, "Pentecostal Student Movement at Howard: 1946–1977," *Spirit* 1, no. 2 (1977), 15.

10. Ibid., 17.

11. Ibid., 21.

12. Reinhold Niebuhr, *Moral Man and Immoral Society* (New York: Charles Scribner's Sons, 1932), 253, 274.

13. Harvey Cox Jr., "Some Personal Reflections on Pentecostalism," *Pneuma, The Journal of the Society for Pentecostal Studies* 15, no. 1 (spring 1993), 33. Cox claims that his course on Pentecostalism was a "roaring success," probably the "most cosmopolitan class meeting anywhere at Harvard," drawing students from Africa, Asia, Latin America, and the United States.

14. Bethune-Cookman College, Eastern Kentucky University, Florida A & M University, Florida State University, Georgia College, Georgia State College, Jackson State University, Kentucky State University, Marshall University, Miles College, Morehead State University, North Carolina A & T University, Ohio State University, Sounds of Blackness (Macalester College), University of Kentucky, University of Louisville, University of North Carolina, University of Southern Florida, Quinnipac College, Wayne State University, and West Virginia Institute of Technology.

15. See Cheryl J. Sanders, "Why We Sing, *Swarthmore College Bulletin* 92, no. 3:64–65.

16. Gilkes, " 'Together and in Harness,' " 687.

17. Katrina Hazzard-Gordon, "Dancing to Rebalance the Universe: African American Secular Dance and Spirituality," *Black Sacred Music* 7, no. 1 (spring 1993), 17.

18. Sounds of Blackness, *Africa to America: The Journey of the Drum*, compact disk recording (Perspective Records, 31454 9006 2, 1994).

19. Gayraud S. Wilmore, introduction to Part I, "The End of an Era: Civil Rights to Black Power," in Gayraud S. Wilmore and James H. Cone, eds., *Black Theology: A Documentary History, 1966–1979* (Maryknoll, N.Y.: Orbis Books, 1979), 15–17.

20. James H. Cone, *Black Theology and Black Power* (New York: Seabury Press, 1969), esp. 55, 110.

21. Wilmore, introduction to Part IV, "Black Theology and the Black Church," in Wilmore and Cone, *Black Theology*, 247.

22. Ibid., 250. See "The New Black Evangelicals" by Ronald C. Potter and "Factors in the Origin and Focus of the National Black Evangelical Association" by William H. Bentley in the same volume.

23. Wilmore, Ibid. in Wilmore and Cone, *Black Theology*, 249.

24. Ronald J. Fowler, "Introduction" in *The Church of God in Black Perspective: Proceedings of the Caucus of Black Churchmen in the Church of God*, April 1970, Cleveland, Ohio (Anderson, Ind.: Shining Light Survey Press, 1970), i.

25. Fowler, *The Church of God*, 5.

26. Ibid., 107.

27. Ibid., 110.

28. James Earl Massey, *Designing the Sermon* (Nashville: Abingdon, 1980); *The Hidden Disciplines* (Anderson, Ind.: Warner Press, 1972); *The Responsible Pulpit* (Anderson, Ind.: Warner Press, 1974); and *The Sermon in Perspective* (Anderson, Ind.: Warner Press, 1976).

29. James Earl Massey, *The Church of God*, 119–120.

30. Benjamin F. Reid, "The National Association Must Live On!" in *The Church of God*, 64–65.

31. Ibid., 66.

32. Ibid.

33. Ibid., 67.

34. Ibid.

35. Ibid., 68.

36. Charles R. Pleasant Sr., "The Black Church's Stand with Reference to Revolution and Rebellion," in *The Church of God*, 105.

37. Peter J. Paris, *The Social Teaching of the Black Churches* (Philadelphia: Fortress Press, 1985), 124–125.

38. "Statement of Mission," National Association of the Church of God, in *1994 Yearbook of the Church of God* (Anderson, Ind.: Leadership Council of the Church of God, 1994), 46.

Chapter 6

1. In 1903, W. E. B. Du Bois used the terms "double-consciousness" and "twoness" to describe the self-understanding of the black American, framing the definitive statement of the dilemma of black racial and cultural identity that has commanded the attention of black intellectuals throughout the twentieth century. See his *The Souls of Black Folk* (New York: New American Library), 214–215. Du Bois's final solution to this dilemma was to leave America and emigrate to Ghana, where he died in 1963.

2. Cornel West's unifying theory of black religion and culture, based on the Marxist interpretations of Antonio Gramsci and Raymond Williams, is essentially grounded in two notions: counterhegemonic culture and the organic intellectual (*Prophesy Deliverance!: An Afro-American Revolutionary Christianity*

[Philadelphia: Westminster Press, 1982], 120–121). See also his essay, "Martin Luther King, Jr.: Prophetic Christian as Organic Intellectual," in *Prophetic Fragments* (Grand Rapids, Mich.: Wm. B. Eerdmans, 1988), 3–12.

3. West, *Prophesy Deliverance!* 69–91.

4. Chancellor Williams, *The Destruction of Black Civilization: Great Issues of a Race from 4500 B.C. to 2000 A.D.* (Chicago, Ill.: Third World Press, 1976), 17.

5. Ibid., 18.

6. Chancellor Williams, "The Socio-Economic Significance of the Store-Front Church Movement in the United States since 1920" (Ph.D. dissertation, American University, 1949; Ann Arbor: University Microfilms, 1970), 219.

7. Williams, *Destruction,* 333.

8. Ibid., 355.

9. Leonard E. Barrett, *Soul-Force: African Heritage in Afro-American Religion* (Garden City, N.Y.: Doubleday, 1974), 3.

10. Ibid., 39.

11. E. Franklin Frazier, *The Negro Church in America* (New York: Schocken Books, 1974), 9.

12. Wilson Jeremiah Moses, "Inventing a Happier Past: National Myth and the Realities of Slavery," in *The Wings of Ethiopia: Studies in African-American Life and Letters* (Ames: Iowa State University Press, 1990), 51.

13. Frazier, *Negro Church,* 9, 122, no. 1. See also Albert J. Raboteau, *Slave Religion* (New York: Oxford University Press, 1978) and Anthony M. Platt, *E. Franklin Frazier Reconsidered* (New Brunswick, N. J.: Rutgers University Press, 1991).

14. Frazier, *Negro Church,* 20.

15. Harold Dean Trulear, "A Critique of Functionalism: Toward a More Holistic Sociology of Afro-American Religion," *The Journal of Religious Thought* 42, no. 1 (spring-summer, 1985): 45–48.

16. Frazier, *Negro Church,* 73.

17. Ibid., 90.

18. Ibid., 77.

19. Platt, *E. Franklin Frazier,* 39.

20. West, *Prophesy Deliverance!* 85.

21. Ibid., 80.

22. Ibid., 84.

23. James Baldwin, *The Fire Next Time* (New York: Dial Press, 1963), 47. Note also James S. Tinney's comment that "Pentecostalism is an important part of James Baldwin's own past (which he tries to exorcise from his haunted memory through his various novels)" in "The Blackness of Pentecostalism," *Spirit* 3, no. 2 (1979), 28.

24. David Leeming, *James Baldwin: A Biography* (New York: Alfred A. Knopf, 1994), 78, 89.

25. Ibid., 375.

26. Ibid., 374.

27. Ibid., 387.

28. West, *Prophesy Deliverance!* 85–86.

29. James H. Cone, *The Spirituals and the Blues: An Interpretation* (New York: Seabury Press, 1972), 1–4.

30. James H. Cone, "Sanctification and Liberation in the Black Religious Tradition," in *Sanctification and Liberation: Liberation Theologies in Light of the Wesleyan Tradition,* ed. Theodore Runyon (Nashville: Abingdon Press, 1981), 185.

31. Ibid., 190.

32. Theophus H. Smith, *Conjuring Culture: Biblical Formations of Black America* (New York: Oxford University Press, 1994), 178.

33. Ibid., 179.

34. C. Eric Lincoln and Lawrence H. Mamiya, *The Black Church in the African American Experience* (Durham, N. C.: Duke University Press, 1990), 179.

35. Howard Thurman, *Deep River and the Negro Spiritual Speaks of Life and Death* (Richmond, Ind.: Friends United Press, 1975), General Introduction, n.p.

36. Ibid., 79.

37. Ibid., 106.

38. Howard Thurman, "Violence and Non-Violence," sermon preached July 14, 1963 on tape in library of the Howard Thurman Educational Trust, cited in Alton B. Pollard III, *Mysticism and Social Change: The Social Witness of Howard Thurman* (New York: Peter Lang, 1992), 112.

39. Ithiel C. Clemmons, James A. Forbes Jr., Bennie Goodwin, and William C. Turner Jr. also made outstanding contributions to Afro-Pentecostal thought during the 1970s.

40. Leonard Lovett, "Black Holiness-Pentecostalism: Implications for Ethics and Social Transformation" (Ph.D. dissertation, 1979 Emory University; Ann Arbor: University Microfilms International), 171.

41. Leonard Lovett, "Perspective on the Black Origins of the Contemporary Pentecostal Movement," *The Journal of the I.T.C.* 1, no. 1, (1973), 48.

42. James S. Tinney, "A Theoretical and Historical Comparison of Black Political and Religious Movements" (Ph.D. dissertation, Howard University, 1978), 278.

43. Tinney, "The Blackness of Pentecostalism," 28.

44. Ibid., 34.

45. Tinney analyzes the ecclesiological and theological implications of homosexual practice among Pentecostals in his "Homosexuality as a Pentecostal Phenomenon," *Spirit* 1, no. 2 (1977), 45–59. See also Charles E. Jones's article, "James Steven Tinney," in *Dictionary of Pentecostal and Charismatic Movements,* ed. Stanley M. Burgess and Gary B. McGee (Grand Rapids, Mich.: Zondervan, 1988), 845–846.

46. Pearl Williams-Jones, "Afro-American Gospel Music: A Crystallization of the Black Aesthetic," *Ethnomusicology* 19, no. 3 (September 1975), 376.

47. Ibid., 384.

48. Pearl Williams-Jones, "Black Pentecostal Women: A Minority Report," *Spirit* 1, no. 2 (1977), 31–44.

49. See Aldon D. Morris, *The Origins of the Civil Rights Movement* (New York: Free Press, 1984) and Jon Michael Spencer's *Protest and Praise* (Minneapolis: Fortress Press, 1990).

50. See Cheryl Townsend Gilkes, "The Role of Women in the Sanctified Church," *Journal of Religious Thought* 43, no. 1 (spring-summer, 1986), 37.

51. Leonard Lovett, "Aspects of the Spiritual Legacy of the Church of God in Christ: Ecumenical Implications," in *Black Witness to the Apostolic Faith*, ed. David T. Shannon and Gayraud S. Wilmore (Grand Rapids, Mich.: Wm. B. Eerdmans, 1985), 41.

Chapter 7

1. W. E. B. Du Bois, *The Souls of Black Folk,* in *Three Negro Classics* (New York: Avon Books, 1965), 215.

2. See Riggins R. Earl, *Dark Symbols, Obscure Signs* (Maryknoll, N.Y.: Orbis Books, 1993), 174; Paul Gilroy, *The Black Atlantic: Modernity and Double Consciousness* (Cambridge: Harvard University Press, 1993); Michael W. Harris, *The Rise of Gospel Blues* (New York: Oxford University Press, 1992), 208; C. Eric Lincoln and Lawrence H. Mamiya, *The Black Church in the African American Experience* (Durham, N. C.: Duke University Press, 1990), 16; Peter J. Paris, *The Social Teaching of the Black Churches* (Philadelphia: Fortress Press, 1985), 28–29; Theophus H. Smith, *Conjuring Culture:Biblical Formations of Black America* (New York: Oxford University Press, 1994), 112; Emilie M. Townes, *Womanist Justice, Womanist Hope* (Atlanta: Scholars Press, 1993), 200–201; Theodore Walker Jr., *Empower the People* (Maryknoll, N.Y.: Orbis Books, 1991), 56; William D. Watley, *Singing the Lord's Song in a Strange Land* (Grand Rapids, Mich.: Wm. B. Eerdmans, 1993), 1–2; and Cornel West, *Race Matters* (Boston: Beacon Press, 1993), 97–98.

3. West, *Race Matters,* 97.

4. Henry Justin Ferry, "Francis James Grimke: Portrait of a Black Puritan" (Ph.D. dissertation, Yale University, 1970), 357.

5. Du Bois, *Souls of Black Folk,* 215.

6. Ephraim Radner, "From 'Liberation' to 'Exile': A New Image for Church Mission," *Christian Century* 106, no. 30 (October 18, 1989), 931.

7. Ibid.

8. Ibid., 932.

9. Ibid., 933.

10. Ibid.

11. Ibid.

12. Ibid.

13. Ibid., 934.

14. Harvey G. Cox Jr., "Some Personal Reflections on Pentecostalism," *Pneuma: The Journal of the Society for Pentecostal Studies* 15, no. 1 (spring 1993), 31.

15. Hans Baer and Merrill Singer, *African-American Religion in the Twentieth Century: Varieties of Protest and Accommodation* (Knoxville: University of Tennessee Press, 1992), 178; and James S. Tinney, "The Prosperity Doctrine: Perverted Economics," *Spirit* 2, no. 1 (1978), 44–45.

16. Roger Finke and Rodney Stark, *The Churching of America, 1776–1990: Winners and Losers in Our Religious Economy* (New Brunswick, N. J.: Rutgers University Press, 1992), 42. See also H. Richard Niebuhr, *The Social Sources of Denominationalism* (New York: Henry Holt, 1929), 20.

17. Wilson Jeremiah Moses, *Black Messiahs and Uncle Toms* (University Park: Pennsylvania State University Press, 1982), 233.

18. Ephraim Radner, "Religious Schooling as Inner-City Ministry," *Christian Century* 108, no. 8 (March 6, 1991), 262.

19. John G. Gammie, *Holiness in Israel* (Minneapolis: Fortress Press, 1989), 195–196. This reference was brought to the author's attention by Gene Rice, Professor of Old Testament at the Howard University School of Divinity.

20. William C. Turner Jr., "Is Pentecostalism Truly Christian?" *Spirit* 2, no. 1 (1978), 38.

21. Ibid., 37.

22. See *The Church of God in Black Perspective* (Anderson, Ind.: Shining Light Survey Press, 1970) and Cheryl J. Sanders, "The Ethics of Holiness and Unity in the Church of God," in *Called to Minister, Empowered to Serve,* ed. Juanita Evans Leonard (Anderson, Ind.: Warner Press, 1989).

23. Frank Bartleman, *How "Pentecost" Came to Los Angeles* (Los Angeles, 1925), 5-43 quoted in Vinson Synan, *The Holiness-Pentecostal Movement in the United States* (Grand Rapids, Mich.: Wm. B. Eerdmans, 1971), 98.

24. "Women Clergy in 21 U. S. Denominations, 1977–1986" from 1988 *Eculink* study, cited in Leonard, *Called to Minister,* 42, 158.

25. Norris Magnuson, *Salvation in the Slums: Evangelical Social Work, 1865–1920* (Metuchen, N.J.: Scarecrow Press, 1977), 41.

26. Ibid., xvi.

27. Ibid., 165–166.

28. Ibid., 178.

29. Ella P. Mitchell and Henry H. Mitchell, "Black Spirituality: The Values in that 'Ol' Time Religion,' " *Journal of the Interdenominational Theological Center* 17, nos. 1 & 2 (fall 1989/spring 1990), 99.

30. Robert M. Franklin, "The Emergence of the Word Churches," unpublished paper presented at the "Pulpit, Pew, and Academy," a conference sponsored by the Interdenominational Theological Center at Callaway Gardens, Georgia, on April 7, 1994.

31. Lillian Ashcraft Webb, *About My Father's Business: The Life of Elder Michaux* (Westport, Conn.: Greenwood Press, 1981), 42.

32. Stephen L. Carter, *The Culture of Disbelief* (New York: Basic Books, 1993), 60.

33. Vincent L. Wimbush, "The Bible and African Americans: An Outline of Interpretive History," in *Stony the Road We Trod: African American Biblical Interpretation,* ed. Cain Hope Felder (Minneapolis: Fortress Press, 1991), 94. See also his "Biblical Historical Study as Liberation: Toward an Afro-Christian Hermeneutic," in *African American Religious Studies,* ed. Gayraud S. Wilmore

(Durham, N.C.: Duke University Press, 1989), originally published in the *Journal of Religious Thought* 42, no. 2 (fall-winter 1985–1986); and " 'Rescue the Perishing': The Importance of Biblical Scholarship in Black Christianity," in *Black Theology: A Documentary History,* Vol. 2: 1980–1992, ed. James H. Cone and Gayraud S. Wilmore (Maryknoll, N.Y.: Orbis Books, 1993), originally published in *Reflection* (Yale Divinity School) 80, no. 2 (1983).

34. Smith, *Conjuring Culture,* 250–256.

35. Leonard E. Barrett, *Soul-Force* (Garden City, N.Y.: Doubleday, 1974), 206–207.

36. John Miller Chernoff, *African Rhythm and African Sensibility: Aesthetics and Social Action in African Musical Idioms* (Chicago: University of Chicago Press, 1979), 157.

37. Chancellor Williams, "The Socio-Economic Significance of the Store-Front Church Movement in the United States Since 1920" (Ph.D. dissertation, American University, 1949; Ann Arbor: University Microfilms, 1970), 219; and *The Destruction of Black Civilization: Great Issues of a Race from 4500 B.C. to 2000 A.D.* (Chicago: Third World Press, 1976), 333. See also the discussion of Williams's book and dissertation in chapter 6.

38. Mary R. Sawyer, *Black Ecumenism: Implementing the Demands of Justice* (Valley Forge, Pa.: Trinity Press International, 1994), 191.

39. The Reverend Eugene Rivers developed the Ten Point Coalition in cooperation with progressive black denominational pastors in Boston and Cambridge to fight crime and gang violence. Rivers's Dorchester home was riddled with bullets two days after the inauguration of his antiviolence community development program, the Boston Freedom Summer '94. See Jim Rice, "Boston Churches Take to the Streets," *Sojourners* 23, no. 7 (August 1994), 12.

40. W. J. Hollenweger, "The Pentecostal Elites and the Pentecostal Poor: A Missed Dialogue?" in *Charismatic Christianity as a Global Culture,* ed. Karla Poewe (Columbia: University of South Carolina Press, 1994), 201.

41. Ibid., 213.

42. Ibid., 207.

43. Ibid., 213.

44. Cheryl Bridges Johns, *Pentecostal Formation: A Pedagogy among the Oppressed* (Sheffield, England: Sheffield Academic Press, 1993), 69.

Conclusion

1. Charles H. Long, *Significations: Signs, Symbols, and Images in the Interpretation of Religion* (Philadelphia: Fortress Press, 1986), 177.

2. Peter J. Paris, *The Social Teaching of the Black Churches* (Philadelphia: Fortress Press, 1985), 84.

3. Cornel West, *Keeping Faith: Philosophy and Race in America* (New York: Routledge, 1993), xiii.

4. See Randall K. Burkett and Richard Newman, eds., *Black Apostles: Afro-American Clergy Confront the Twentieth Century* (Boston: G. K. Hall, 1978).

5. Paul Gilroy, *The Black Atlantic* (Cambridge: Harvard University Press, 1993), 24.

6. Theophus H. Smith, *Conjuring Culture: Biblical Formation of Black America* (New York: Oxford University Press, 1994), 128. See also Elias Farajajé-Jones, *In Search of Zion: The Spiritual Significance of Africa in Black Religious Movements* (Bern: Peter Lang, 1990).

7. Anthony M. Platt, *E. Franklin Frazier Reconsidered* (New Brunswick, N.J.: Rutgers University Press, 1991), 220. The bulk of Frazier's papers are deposited in the Moorland-Spingarn Research Center at Howard University.

8. Ibid., 221.

9. West, *Keeping Faith,* x, xvi.

10. Ibid., xiv.

11. bell hooks, *Yearning: Race, Gender, and Cultural Politics* (Boston: South End Press, 1990), 41–43.

12. See chapter 6, also David Leeming's discussion of Baldwin's exile and "welcome table" in France in *James Baldwin: A Biography* (New York: Alfred A. Knopf, 1994), 374.

13. Farajajé-Jones, *In Search of Zion,* 34.

14. Walter Brueggemann, *Hopeful Imagination: Prophetic Voices in Exile* (Philadelphia: Fortress Press, 1986), 107.

15. William Herbert Brewster Sr., "Rememberings," in *We'll Understand It Better By and By,* ed. Bernice Johnson Reagon (Washington, D.C.: Smithsonian Institution Press, 1992), 193.

16. Anthony Heilbut, " 'If I Fail, You Tell the World I Tried': William Herbert Brewster on Records," in Reagon, *We'll Understand It,* 234.

17. See Cheryl Townsend Gilkes, " 'Mother to the Motherless, Father to the Fatherless:' Power, Gender, and Community in an Afrocentric Biblical Tradition," *Semeia* 47 (1989), 59–60. One of the most memorable stanzas of the song describes God as a "mother to the motherless" and a "father to the fatherless."

18. Brueggemann, *Hopeful Imagination,* 107.

19. Jeremiah A. Wright Jr., *What Makes You So Strong?* ed. Jini Kilgore Ross (Valley Forge, Pa.: Judson Press, 1993), 138.

20. Charles M. Wiltse, ed., *David Walker's Appeal to the Coloured Citizens of the World* (New York: Hill and Wang, 1965), 70.

21. Maria W. Steward, "Religion and the Pure Principles of Morality," in *Early Negro Writing,* ed. Dorothy Porter (Boston: Beacon Press, 1971), 469–470.

22. See Cornel West, "Nihilism in Black America" in *Race Matters* (Boston: Beacon Press, 1993),9–20. See also Frazier's classic statement of the crisis of the black middle class in *Black Bourgeoisie* (New York: Free Press, 1957), 26: "Because of their social isolation and lack of a cultural tradition, the members of the black bourgeoisie in the United States seem to be in the process of becoming NOBODY."

23. Brueggemann, *Hopeful Imagination,* 129.

24. Ibid., 112.

25. Alice Walker, *In Search of Our Mothers' Gardens,* (New York: Harcourt, Brace Jovanovich, 1983), 5 quoted in ibid., 130.

26. Brewster, "Rememberings," 201.

27. Michael W. Harris, *The Rise of Gospel Blues: The Music of Thomas Andrew Dorsey in the Urban Church* (New York: Oxford University Press, 1992), 258–259.

28. Brueggemann, *Hopeful Imagination,* 130. In the text, Brueggemann paraphrases Dorsey's gospel song.

29. Harvey Cox, *Fire from Heaven: The Rise of Pentecostal Spirituality and the Reshaping of Religion in the Twenty-first Century* (Reading, Mass.: Addison-Wesley, 1995), 320–321.

30. For a comprehensive scholarly overview of the Pentecostal and charismatic churches in global perspective, see Karla Poewe, ed., *Charismatic Christianity as a Global Culture* (Columbia: University of South Carolina Press, 1994).

Index